THERMOPLASTIC ELASTOMERS - SYNTHESIS AND APPLICATIONS

Edited by **Chapal Kumar Das**

Thermoplastic Elastomers - Synthesis and Applications
http://dx.doi.org/10.5772/59647
Edited by Chapal Kumar Das

Contributors

Lavinia Ardelean, Cristina Maria Bortun, Angela Codruta Podariu, Laura Cristina Rusu, Maria Rutkowska, Katarzyna Krasowska, Heimowska Aleksandra, Pierre Carreau, Brzeska Joanna, Joanna Rokicka, Ryszard Ukielski, Jaroslav Mosnacek, Marketa Ilcikova, Miroslav Mrlik, Tanya Das

Published by InTech

Janeza Trdine 9, 51000 Rijeka, Croatia

Publishing Process Manager Iva Lipovic

Technical Editor InTech DTP team

Cover InTech Design team

First published November, 2015

Printed in Croatia

Additional hard copies can be obtained from orders@intechopen.com

Thermoplastic Elastomers - Synthesis and Applications, Edited by Chapal Kumar Das

p. cm.

ISBN 978-953-51-2223-4

For EU product safety concerns:
IN TECH d.o.o., Prolaz Marije Krucifikse Kozulić 3, 51000 Rijeka, Croatia,
info@intechopen.com or visit our website at intechopen.com.

Meet the editor

Professor Chapal Kumar Das is a professor at the Materials Science Centre, IIT Kharagpur, India. He has received his Ph.D. from the same institute. He has both industrial and academic experience. His research interest lies in the fields of polymer blends and alloys, high-performance composites based on LCP, self-reinforcing elastomers, nano-polymer composites, devulcanization of scrap tires, direct fluorination of plastics, flexible engineering composites for defense applications, short Kevlar fiber composites based on thermoplastics, carbon nanotube–polymer composites, modification of nanofillers, welding of thermoplastic nanocomposites, and elastomeric thin films. His present research interest is in the development of supercapacitors, high-power microwave-absorbing materials, and fuel and solar cells. He has contributed 15 scientific chapters in books and encyclopedias. He has completed 17 high-value projects. He has supervised 32 Ph.D. students and 36 M.Tech./M.S. students. He has published about 370 papers in international journals and about 40 papers in national journals. He has travelled widely throughout various countries for different collaborative projects and international conferences. He served as the Head of the Materials Science Centre, IIT Kharagpur. He is the recipient of the Lady Davis fellowship, Israel. He is a fellow of IRI and a life member of MRSI and the Polymer Society.

Contents

Preface

Thermoplastic elastomers (TPEs), sometimes referred to as thermoplastic rubbers, are a class of copolymers or a physical mix of polymers (usually a plastic part and a rubbery part) which consist of materials with both thermoplastic and elastomeric properties. A TPE is a rubbery material with properties and functional performance similar to those of conventional vulcanized rubber at ambient temperature; yet it can be processed in a molten state as a thermoplastic polymer at an elevated temperature. Most of the elastomers are thermosets; thermoplastics are in contrast relatively easy to manufacture, for example, by injection molding. A TPE shows advantages typical of both rubbery materials and plastic materials. The principal difference between thermoset elastomers and thermoplastic elastomers is in the type of the crosslinking bond in their structures. In fact, crosslinking is a critical structural factor which contributes to imparting high elastic properties. TPEs are a class of polymers bridging between the service properties of elastomers and the processing properties of thermoplastics. Thus, TPEs combine the conventional rubbers and thermoplastics. They provide the simplest way of achieving outstanding properties by simply varying the blend compositions and the viscosity of the components and compounding ingredients at a low cost.

A thermoplastic elastomer must fulfill the following three essential characteristics:

The ability to be stretched to moderate elongations and, upon the removal of stress, return to something close to its original shape

Processable as a melt at elevated temperature

Absence of significant creep

TPE materials have the potential to be recyclable since they can be molded, extruded, and reused like plastics, but they have the typical elastic properties of rubbers which are not recyclable owing to their thermosetting characteristics. TPEs also require little or no compounding and do not require the addition of reinforcing agents, stabilizers, or cure systems. Hence, batch-to-batch variations in weighting and metering components are absent, leading to improved consistency in both raw materials and fabricated articles. TPEs can be easily colored by most types of dyes, and more economical control of product quality is possible.

The applications of TPEs are listed below:

Gaskets, Seals, Stoppers, Housings, Rollers, Valves, Bumpers, Wheels, Pads, Casters

Strain Reliefs, Cable Jacketing, Switch Contact Points, Bellows, Fuel Line Covers, Housings

Shoe Soles/Heels, Wrist Straps, Cushions, Airbag Doors

Keycaps, Cosmetic Cases, Handles, Grips, Push Buttons, Knobs

Professor Chapal Kumar Das

Materials Science Centre

IIT Kharagpur, India

Chapter 1

Synthesis and Properties of Polyurethanes Based on Synthetic Polyhydroxybutyrate for Medical Application

Joanna Brzeska

Additional information is available at the end of the chapter

http://dx.doi.org/10.5772/60933

Abstract

Polyurethanes is a group of polymers whose unique properties make them useful in both the construction and the textile industry, and even in tissue engineering. One small, but very significant, urethane group connects specially selected macrochains to obtain a material with established properties.

The chapter is a literature review for research on the synthesis and properties of new degradable polyurethanes that contain soft segments synthesized with synthetic telechelic poly([R,S]-3-hydroxybutyrate) (R,S-PHB). Incorporation of oligomeric R,S-PHB (what was found degradable and biocompatible) into polyurethane structures gives them a chance of improving the properties important for medical applications. Aliphatic and aromatic polyurethanes with different soft segments were investigated due to their potential to be used in soft tissue regeneration.

Keywords: Polyurethanes, synthetic polyhydroxybutyrate, medical application

1. Introduction

Polyurethane (PUR) is a large and very diverse group of polymers, including elastomeric and thermoplastic materials (liquid, millable), foams, and ionomers in the aqueous dispersions [1–3]. The preparation of polyurethanes in such various forms has allowed their broad use in such industries as construction, engineering, automotive, textile, and medical. In medicine, their biocompatible, biostatic, and biodegradable properties are very desirable. The required properties can be achieved using appropriate monomers for polyurethane synthesis and for preparing their composites.

1.1. Polyurethanes in medicine

PURs' career in medicine has been going for almost 60 years since polyurethane foam for breast implants has been patented [4, 5]. According to studies of Jose Abel de la Pen˜a-Salcedo et al. [6] performed at the Institute for Plastic Surgery, polyurethane-covered implants are still the best option for breast reconstruction.

PURs are already used or investigated for utilization as membranes for wound dressing [7], as meniscal scaffold to treat partial meniscal loss [8], as drug nanocarriers for endovascular applications [9], as controlled release membrane system for delivery of ketoprofen [10], or as the biostable polyurethane/hydroxyapatite composites for bone replacement materials [11]. Whereas the shape memory PUR (based on PCL) used as wire in orthodontic appliance, could effectively align the teeth [12]. Waterborne polyurethane with chitosan as chain extender was also studied as antimicrobial agent for acrylic fabrics that could be used for the manufacture of blankets and carpets in hospitals [13].

One of the most important uses of PURs in medicine is the preparation of the implants for cardiovascular diseases, the use of which the specific properties of PUR (high mechanical strength, toughness and flexibility without the addition of modifiers, and good hemocompatibility resulting from occurring on the surface hydrophilic-hydrophobic balance) are very important.

1.2. Polyurethanes biostatic and biocidic

The very important danger during implantation is connected with bacterial and fungi contamination. Aside from antibiotics treatment after surgery, the use of biostatic implant is essential for the success implantation.

The bacterial adherence and encrustation was reduced after the immobilization on PUR (Tecoflex®) surface by polyvinylpyrrolidone-iodine (PVP-I) complex [14]. Modified films were much more hydrophilic than original films.

Surfaces coated with quaternized PUR possessed the antibacterial and antiviral properties [15]. For wound dressing application, the asymmetric PUR membrane, with diamino containing antibiotic sulphanilamide used as a chain extender could be utilized [16]. These antibiotic-conjugated PUR are enzymatic susceptible what fit them antibacterial activity. The antibacterial properties can also be achieved by using nanoparticles for polyurethane or its composites obtaining. The nanosilver nanoparticles are very often used [17].

1.3. Biodegradable polyurethanes

The first applications of polymers required their high resistance to environmental factors. They had to be stable (non-degradable) under the operating conditions throughout. This applies to the construction, packaging, textile, mechanical, medical, as well as other industries. However, the growing environmental burden of polymer waste has caused plastics to be gradually replaced by degradable materials. In medicine, dynamically developing tissue engineering suggests many interesting solutions with biodegradable polymeric materials.

PURs were originally considered to be very resistant to environmental impacts and were used in the construction of the first "artificial hearts". So why not produce something biocompatible as polyurethane and simultaneously degradable for tissue reconstruction? Since the first time this question was formulated, a lot of really interesting investigations were conducted.

The use of monomers susceptible to environmental factors allows the production of biodegradable PURs potentially suitable to create scaffolds for the growth of living cells or as temporary implants. Hydrolysis-sensitive ester groups, mostly with oligomerols, are introduced into the PUR structure to build the soft segments.

Wang and co-workers [18] synthesized biodegradable polyurethanes with 11,11′-dithiodiundecanol employed as a soft segment. They concluded that the molecular weight of PURs substantially decreased and the surface morphology was significantly eroded after 8 days of incubation in SBF with reduced glutathione. Gisele Rodrigues da Silva et al. [19] observed that the biodegradable PURs based on PCL were able to release dexamethasone acetate for 371 days at almost constant rates.

1.4. Polyurethanes that are more biocompatible

Using of natural components (or their synthetic substitutes) for PURs synthesis is one of method for making them more biocompatible. Saralegi et al. used castor oil for soft segment building [20]. They obtained the shape memory thermoplastic PUR due to the addition of cellulose nanocrystals. L-arginine, glycine, and L-aspartic acid were used as chain extenders in poly(urea)urethanes synthesis by the Chan-Chan group [21]. The authors concluded that PURs containing L-arginine would be a potential candidate for cardiovascular applications and angiogenesis. Studies of Lin Jia and co-workers [22] showed the lack of cytotoxicity of PUR/collagen and PUR/gelatin nanofibrous scaffolds. The authors indicated on sufficient mechanical properties, supported SMC proliferation, and assisted in oriented morphological alignment of cells of PURs with L-arginine that make them the appropriate candidate for vascular tissue engineering.

Very important for obtaining biocompatible material is the use for their synthesis substrates that are non-toxic and degraded into non-toxic compounds. In medical applications, 4,4'-methylene dicyclohexyl diisocyanate (H_{12}MDI) successfully replaced 4,4'-diphenylmethane diisocyanate (MDI), especially in the synthesis of biodegradable materials, thereby reducing the risk of creation of carcinogenic aromatic diamine as degradation product of PUR based on MDI.

Using natural components for polyurethanes building very often gives them a chance to be biocompatible and degradable simultaneously. The important groups of substrates useful for polyurethanes synthesis are polyhydroxyacids (PHA). Among them, polyhydroxybutyrate (PHB) is the most often used.

1.5. Biosynthesized polyhydroxybutyrate

As it was mentioned before, the most popular polyhydroxyacid is PHB. Since the 1920s when Lemoigne found the bacterial granules of supplementary material (later called polyhydroxybutyrate) in *Bacillus megaterium*, intensive researches on the biological and chemical obtaining of PHB, its properties, and application were conducted. The natural origin, biodegradability, and biocompatibility of polyhydroxybutyrate made it such interesting material for medical applications whereas its low water vapor permeability, which is close to that low-density polyethylene, promote it to food packaging applications [23].

PHB degrades into 3-hydroxybutyric acid, a common metabolite in human blood. 3-hydroxybutyric acid is produced in ketone bodies of mammals during the prolonged starvation [24]. 3-hydroxybutyric acid belongs to short-chain fatty acids and reveals antibacterial activity [25].

A lot of investigations suggest that PHB is non-genotoxic [26]. All this features promote PHB for medical applications. Lee and co-workers prepared a carrier system with targeting capability for imaging and drug delivery to cancer cells using catalytic characteristics of PHA synthase [27]. They found an attractive way of preparing functionalized nanoparticles by effective coupling between the hydrophobic surface of PHB nanoparticle and PHB chain grown from the fusion enzyme. Medvecký and co-workers [28] found that the addition of hydrophobic PHB microparticles into the calcium phosphate cement significantly improves the initial cement properties (the higher tensile and compressive strengths) and makes it a very promising material for bone substituting. Another way of PHB utilization in tissue reconstruction is by using it as the PHB–chitosan biopolymer scaffolds [29], PHB-calcium phosphate/chitosan barrier membrane [30], hydroxyapatite/PHB composites [31], or biodegradable stents [32]. The investigations of Shishatskaya at el. [33, 34] indicate that PHB is a good candidate for fabricating prolonged-action drugs as microparticles intended for intramuscular injection. Whereas Althuri and co-workers [35] concluded that folate functionalized PHB nanoparticles can be used as a polymer matrix to carry toxic drug compounds to targeted sites for treatment of life-threatening diseases such as cancer.

Nonetheless, the inherent brittleness and stiffness (connected to its semicrystalline nature) and inferior thermal stability, in addition to relative high cost, have blocked the popular use of PHB.

It is known that even the short exposure of PHB to temperatures near 180°C degrades it to olefinic and carboxylic acid compounds (e.g., crotonic acid) and various oligomers. Also, during storage, the degree of crystallization of polymer increases and causes the formation of irregular pores on its material surface and causes even higher stiffness. These disadvantages can be reduced by mixing of PHB with plasticizers, such as low molecular weight PHB [36], carboxyl-terminated butadiene acrylonitrile rubber, or biocompatible polyvinylpyrrolidone polymeric additives [37].

1.6. Chemically synthesized poly([R,S]-3-hydroxybutyrate)

The chemically synthesized substitute of natural PHB is synthetic poly([R,S]-3-hydroxybutyrate) (R,S-PHB). Synthetic R,S-PHB can be obtained by anionic ring-opening polymerization of (R,S)-ß-butyrolactone. The supramolecular acid sodium salt complex of 3-hydroxybutyric acid ether 18-crown-6 can then be used as the initiator. The polymerization process is carried out in THF at room temperature. The resulting polymer could be reacted with 2-bromo- or 2-iodoethanol, finally causing PHB to be terminated with hydroxy groups on both sides [38–41].

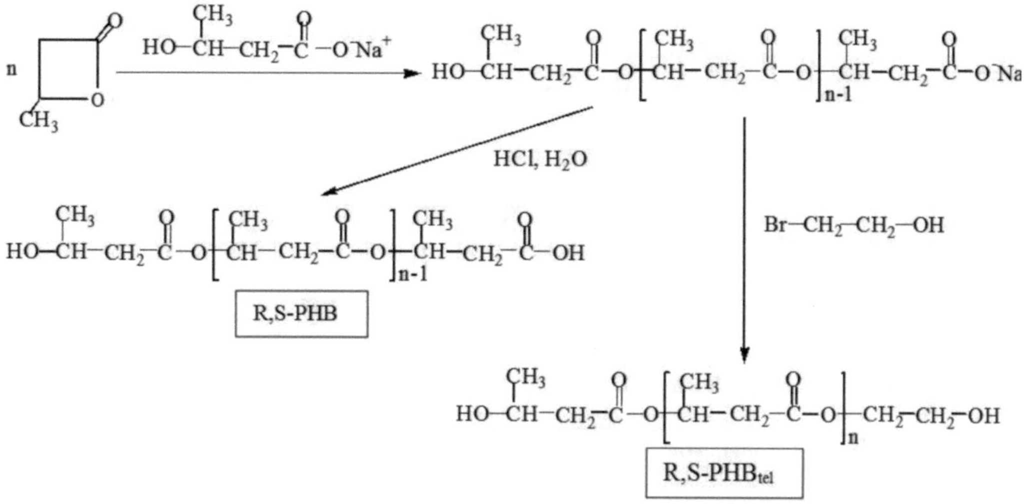

Figure 1. Scheme of how telechelic R,S-PHB (OH-terminated) and R,S-PHB (OH- and COOH-terminated) are obtained.

The literature indicated that materials obtained with synthetic R,S-PHB was biocompatible and biodegradable. The degradation products of the temporary patch made from PHB/R,S-PHB blends were metabolized and did not evoke inflammatory reactions [42]. Freier et al. [43] found that after 26 weeks of implantation of the patches (made with PHB/R,S-PHB blends) in the abdomen of rats, the loss of intestines of animals was almost completely restored and the introduced material had been substantially degraded.

Piddubnyak et al. [40] conducted a series of studies confirming the biocompatibility and non-toxicity of synthetic [R,S]-3-hydroxybutyrate oligomerols. The possibility of their formation into spherical particles of a diameter <1 micron, suggested that they could be used in obtaining nonsteroidal anti-inflammatory drugs. In the form of an aqueous dispersion, they could be introduced into the body through intravenous, intramuscular, or subcutaneous administration [44].

1.7. Marriage of PUR and PHB advantages in one product

Using for PUR synthesis, almost completely amorphous R,S-PHB that is close to its original state in the cell, ought to be utilized to obtain biocompatible and biodegradable material useful for medical application.

The work is a review of the research on the synthesis and properties of PURs containing synthetic poly([R,S]-3-hydroxybutyrate) and polycaprolactonediol or polyoxytetramethylenediol in soft segments in the structure in terms of applications as medical devices. The properties of polyurethanes that could determine their usefulness for medical application (structure, morphology of surface, thermal and mechanical properties, water and oil sorption, density, degradability, spinnability, compatibility, and biostatic properties) were estimated.

Works with PURs based on natural and synthetic PHB are collected in Table 1.

Kind of PHB origin	Characteristic	Ref.
bacterial	Higher crystallinity than PURs without PHB.	[45]
bacterial (as copolymer P3/4HB)	High molecular weight and narrow molecular weight distribution.	[46]
bacterial (as copolymer P3/4HB)	Narrow distribution and suitable crystallinity to prepare films and pads. Non-toxic for cell growth and proliferation.	[47]
bacterial	Degradability with creation of 3-hydroxybutyric acid and crotonic acid as degradation products.	[48]
synthetic	Hydrolytic degradation of PURs based on PCL/HB increased with increasing of PHB fraction.	[49]
synthetic	The way of PURs based on R,S-PHB obtaining.	[50]
synthetic	Presence of R,S-PHB influenced on the structure of PURs.	[51]
synthetic	Degradability of PURs increased after using of R,S-PHB for their building.	[52]
synthetic	Electrospinning of PURs based on R,S-PHB.	[53]
synthetic	Degradability of PURs increased after using of R,S-PHB for their building.	[54]

Table 1. Polyurethanes based on PHB.

2. Experimental

2.1. Materials

Oligoesters, such as polycaprolactonediol (PCL) and polyhydroxybutyratediol (PHB), are used in the synthesis of polyesterurethanes. The oxidative-sensitive ether groups can be introduced into polyurethanes with polyoxytetramethylenediol (PTMG). 4,4'-diphenylmethane diisocyanate (MDI) and 4,4'-methylene dicyclohexyl diisocyanate (H_{12}MDI) are isocyanates that are often used for building of hard segments of polyurethanes.

Aromatic and aliphatic polyurethanes with synthetic poly([R,S]-3-hydroxybutyrate) incorporated into the soft segments structure were obtained and investigated.

2.1.1. Materials for polyurethanes synthesis

- Before the synthesis of polyurethanes, R,S-PHB (Mn~2000) (CMPW, PAN Zabrze), PCL (Mn~2000) (Aldrich), and PTMG (Mn~2000) (Aldrich) were dried by heating at 60°C–90°C for 3 h under reduced pressure;
- 4,4'-diphenylmethane diisocyanate MDI (Aldrich) was filtered and melted at temperature 40°C;

- 4,4'-methylene dicyclohexyl diisocyanate ($H_{12}MDI$) (Alfa Aesar) (mixture of isomers) was purified via vacuum distillation;
- the chain extender 1,4-butanediol (1,4-BD) (Aldrich) was freed from moisture by an azeotropic distillation with benzene prior to use;
- the solvent dimethylformamide (DMF) (Labscan Ltd) was dehydrated over P_2O_5 and distilled under low pressure before synthesis; and
- catalysts dibutyltindilaurate (DBTDL) (Akra Chem) and stannous octoate (OSn) (Akra Chem) were used as received. Stannous octoate was approved by the US Food and Drug Administration as the catalyst for polyurethane synthesis [55].

2.1.2. Polyurethane synthesis and sample preparation

The synthesis of polyurethanes was carried out in a two-step reaction at the vacuum reactor, as previously described [56].

First, the prepolymer was prepared from oligomerols and $H_{12}MDI$ or MDI, at 60°C–90°C in a presence of a catalyst at reduced pressure according to the appropriate required molar ratio of NCO:OH groups for 2–3 h. Oligomerols used in synthesis: a mixture of PCL and R,S-PHB or a mixture of PTMG and R,S-PHB. For comparison, PURs based only on PCL or PTMG without R,S-PHB were also obtained. The synthesis of prepolymer was carried on mass but next the prepolymer was dissolved in DMF to solid mass concentration of 40%. The chain extender (1,4-BD) was added to obtain equimolar ratio NCO:OH groups. The propagation reaction of prepolymer was carried on for 2–3 h at 60°C.

PUR	**Molar ratio of OH groups of reagents used in PURs synthesis**				**Molar ratio of NCO/OH in prepolymer (diisocyanate)**
	PCL	R,S-PHB	PTMG	1.4-BD	
PUR-2A-1	-	0.23	0.77	2.7	3.7 (MDI)
PUR-2A-2	-	-	1	2.7	3.7 (MDI)
PUR-2A-3	0.77	0.23	-	2.7	3.7 (MDI)
PUR-2A-4	1	-	-	2.7	3.7 (MDI)
PUR-3A-1	-	0.23	0.77	2.7	3.7 ($H_{12}MDI$)
PUR-3A-2	-	-	1	2.7	3.7 ($H_{12}MDI$)
PUR-3A-3	0.77	0.23	-	2.7	3.7 ($H_{12}MDI$)
PUR-3A-4	1	-	-	2.7	3.7 ($H_{12}MDI$)

Table 2. Composition of the obtained polyurethanes.

After the extension of prepolymer chains, the solution of polyurethane was poured on Teflon plates and heated for solvent evaporating (2 h/80°C). Next, the foils were heated in a vacuum dryer for reaction completing (5 h/105°C). Before the estimation of polyurethanes properties, the foils were conditioned at room temperature at least 2 weeks.

The obtained and investigated PURs differed in soft and hard segment structures and in their ratio (Table 2).

2.2. Methods of investigations, obtained results, and discussion

2.2.1. The structure of obtained polyurethanes

The structures of obtained aromatic and aliphatic PURs were investigated using FT-IR and ^{1}HNMR methods (results presented in [51]).

The value of vibration absorption of the carbonyl group in the ester moiety at 1,740 cm^{-1} is indicative of the presence of the amorphous phase of polyhydroxybutyrate (the presence of the crystal phase of stretching vibration of C=O would be observed at 1,725 cm^{-1}) [56, 57]. These differences in the frequencies corresponding to the vibrations of the carbonyl bond were explained by Wu and co-workers [58] by the decrease in oxygen dipole moment under the influence of hydrogen from a neighboring chain. The interaction is stronger when the oxygen is closer to the hydrogen atom. Amorphousness of R,S-PHB used in the synthesis of polyurethane, was also confirmed by the presence of bands of CH_3 at 2,985 cm^{-1} and C-O-C stretching vibration band at 1,186 cm^{-1} [57, 59]. It is known that asymmetric stretching vibration of CH_3 at 3,009 cm^{-1}, 2,995 cm^{-1}, 2,974 cm^{-1}, and 2,967 cm^{-1} in the natural polyhydroxybutyrate indicate its crystallinity [57].

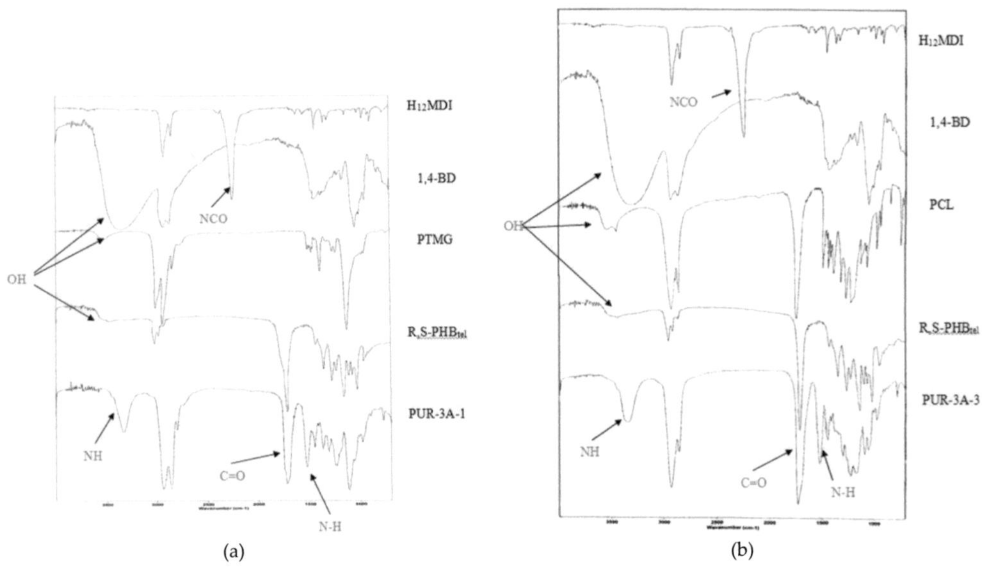

Figure 2. FT-IR spectra of substrates and PUR-3A-1 (a) and substrates and PUR-3A-3 (b).

FT-IR spectra supported the formation of urethane groups during polyaddition reaction of OH groups of oligomerols (R,S-PHB, PCL, and PTMG) and NCO groups of diisocyanates (MDI and H_{12}MDI).

In the spectra of PUR-2 and PUR-3 series with PTMG in soft segments (PUR-2A-2, PUR-2B-2, PUR-3A-2 and PUR-3B-2), a single band (without shoulder) at around 3,330 cm^{-1} corresponded to the NH stretching vibration of urethane groups. The absence of any bands above 3,500 cm^{-1} indicated that all group -NH were involved in the formation of hydrogen bonds.

Simultaneously, in an area corresponding to the stretching vibration of C=O groups of the urethane two bands (at 1,720 cm^{-1}/1,705 cm^{-1} for PUR-3 and at around 1,732 cm^{-1}/1,703 cm^{-1} for PUR-2) were observed. The intensities of these bands were comparable for PUR-3A-2. The literature indicated that the first band corresponded to vibration of free groups C=O and the other for associated groups [60]. The presence of free groups C=O suggested that not all -NH were involved in the formation of hydrogen bonds within the urethane groups, but some of them could be associated with the oxygen moiety of ether. The intensity of the absorption bands of the C=O hydrogen bonded clearly decreased after the introduction of R,S-PHB into the soft segments. On spectra PUR-3A-1 at 1,740 cm^{-1} appeared the band of stretching vibration characteristic for non-hydrogen bonded C=O of ester groups.

In the spectra of all polyurethanes with PCL in their structure, there were clear bands of stretching vibration characteristic for associated -NH. For polymers PUR-2 they were at 3,334 cm^{-1} and at about 3,360 cm^{-1} for PUR-3. At frequencies a bit lower, the presence of broad bands with low intensity was also observed. It indicates the existence of two types of hydrogen bonds, between the -NH groups and the carbonyl group of urethane and between -NH and ester group. On the other hand, the intensity of the bands of stretching, hydrogen-bonded C=O of urethane (amide I band) and ester groups on PUR-2A-3 and PUR-2A-4 spectra (at 1,702 cm^{-1} and 1,705 cm^{-1}) was small.

Thus, the FT-IR spectra of the polyurethanes indicated that urethane groups (partly hydrogen bonded) were formed and the end groups of substrates were completely converted.

On ^{1}HNMR spectra of aliphatic PURs, peaks due to the NH group were observed in the region of 6.6–7.1 ppm [51], whereas on the spectra of aromatic PUR they were located at 9.2–9.5 ppm. The NH groups coming from urethane groups were forming hydrogen bonds with carbonyl groups of ester and urethane groups and the hydrogen bonds with ether groups. It was also concluded that the greater NCO:OH ratio, the more urethane-urethane hydrogen bonds were formed [51]. An addition, R,S-PHB caused the slight increase in number of urethane-urethane hydrogen bonds. The presence of an allophanate structures was observed at 8.5 ppm (for aromatic PURs) and as a very small peak at 9.65 ppm (for aliphatic polyurethanes based on PTMG).

2.2.2. Thermal properties

The presence of the synthetic R,S-PHB (with a lateral methyl group) in soft segments caused a disturbance in the phase separation and increase the glass transition temperature (estimated by DSC at a heating rate of 20°/min, at a temperature ranging from - 80°C to + 200°C) of the aromatic and aliphatic PURs in comparison to PURs without R,S-PHB [61,62].

sample	T_g [°C]	Tm1 [°C]	ΔH_1 [J/g]	T_{m2} [°C]	ΔH_2 [J/g]
PTMG	-70.8	47.8	129.6	-	-
PCL	-60.8	68.3	84.5	-	-
R,S-PHB	-12.3	56.3	4.3	-	-
PUR-2A-1	-67.6	-	-	174.1	1.3
PUR-2A-2	-73.3	35.0	2.0	177.0	12.6
PUR-2A-3	-28.5	44.0	1.4	175.2	7.3
PUR-2A-4	-38.5	51.3	2.8	179.1	5.7
PUR-3A-1	-55.3; -10.5	37.8	0.7	107.4; 135.9; 157.5	2.5; 4.0; 2.2
PUR-3A-2	-55.3	39.8	0.8	164.5	17.0
PUR-3A-3	-32.4	50.7	3.6	90.2; 138.1	4.7; 9.1
PUR-3A-4	-49.1	50.2	19.5	107.7; 131.6; 143.0	5.0; 5.1; 2.9

Table 3. Thermal properties of oligomerols and polyurethanes (results presented in [61, 62]).

Incorporation of oligomerols into PUR structures generally increased their glass temperatures. Moreover, using R,S-PHB for soft segment building increased the Tg of soft segments.

Obtained PURs were amorphous with the low value of crystalline phase. The introduction of R,S-PHB into PUR structures generally reduced crystallinity of soft segments (lower melting enthalpy was observed) [53]. In particular, it caused the reduced susceptibility of PCL to crystallization, which was the result of partial miscibility of both oligoestroles [62].

Temperatures in the range 90°C–179°C indicated the presence of long-range order of hard segments. A few melting endotherms on DSC thermograms (in the mentioned range) of aliphatic PURs (series PUR-3) were probably the result of using nonlinear diisocyanate for hard segment building. H_{12}MDI was a mixture of stereoisomers that may be formed into crystallites with different construction and size, and polymorphism.

The low degree of crystallinity suggested that investigated PURs could be degradable under the conditions of a living body.

2.2.3. Microscopic observation

The surfaces of the samples and surfaces revealed after breaking of obtained PUR samples in liquid nitrogen (cryogenically broken samples), they were tested by Transmission Electron Microscopy (TEM) using a two-step replica.

Studies of electron microscopy of PUR samples showed that they were characterized by varying degrees of homogeneity of the physical (morphological), depending on the chemical composition of the polymers. For samples with reported crystallization of soft segments, crystalline elements were usually in the form of spherulites.

The surface of aromatic polyurethane PUR-2A series samples was smooth or only slightly rough. The cryogenically fractured surface of these samples were homogeneous in the micrometer scale, which indicated that the cracking (during the preparation of the samples) occurred between the crystalline areas, and thus in the weaker points of the polymers. The observed morphology was characteristic of non-crosslinked polymers. It was called un-radial or "mount-depression" morphology [63].

The aliphatic PUR surfaces and their cryogenically fractured surfaces were a bit rougher than the aromatic ones and that could influence their degradability. The surface morphology was comparable for all investigated samples.

There were no significant differences in the morphology of the samples of PUR based on PTMG or PCL. It was found, however, that the introduction of poly([R,S]-3-hydroxybutyrate) into the soft segment structures resulted in a slight decrease in the roughness of the surface of test samples and that could favorably influence the adhesion of blood elements in the study of the effects on blood parameters of polymers [64].

2.2.4. Mechanical properties

The mechanical properties determined for the obtained PUR materials included hardness (in degrees Shore A), tensile strength (Rr), and elongation at break (εr).

PUR	Hardness [°Shore A]	R_r [MPa]	$R_{r\,ster.}$ [MPa]	εr [%]	$\varepsilon_{rster.}$ [%]
PUR-2A-1	82	3.5±0.1	4.6±0.3	110±10	127±6
PUR-2A-2	85	24.6±0.5	20.2±1.8	492±11	734±67
PUR-2A-3	81	9.8±0.3	7.8±0.8	532±40	537±18
PUR-2A-4	82	15.4±0.6	11.1±1.8	616±53	685±139
PUR-3A-1	80	6.9±1.9	12.0±1.2	183±81	403±74
PUR-3A-2	76	7.9±0.5	5.7±0.8	383±37	143±74
PUR-3A-3	86	8.3±1.5	9.4±1.5	361±80	287±99
PUR-3A-4	84	9.0±0.5	8.2±2.3	30±11	29±6

Table 4. The hardness and tensile strength (±standard deviation) before and after the sterilization of PURs (results partially presented in [61,62]).

It was concluded that incorporation of R,S-PHB into aromatic and aliphatic PUR structures slightly reduced their tensile strength and elongation. The hardness of polyurethanes was in the range of commercial polyurethane elastomers used in medicine [61, 62].

The tensile strength of PURs obtained with the participation of the aliphatic diisocyanate (H_{12}MDI) was generally lower than PURs containing asymmetric and aromatic MDI. Investigations of model PURs (constructed only with H_{12}MDI and PTMG, without chain extenders), indicated that in these polymers disordered hydrogen bonding were formed, making the phase separation of soft and hard segments difficult what reduced the mechanical strength of the materials [60].

The influence of sample sterilization on their mechanical properties was estimated. Gas plasma technology (dihydrogen peroxide) was used for sterilization. In some cases, tensile strength ($Rr_{ster.}$) and elongation at the break ($\varepsilon r_{ster.}$) of samples increased after the sterilization process. During the plasma sterilization, the free radicals were generated that could lead to slight cross-linking chains, thereby increasing the elasticity and tensile strength of the obtained PURs [61, 62].

2.2.5. Density

Easy penetration of water and degrading factors into the material is an important factor in the degradation of polymers. The density of the material is one of the parameters that determine the sorption of water.

PUR	Density ±SD [g/cm³]	PUR	Density ±SD [g/cm³]
PUR-2A-1	1.098±0.024	PUR-3A-1	1.081±0.002
PUR-2A-2	1.089±0.004	PUR-3A-2	1.051±0.004
PUR-2A-3	1.189±0.007	PUR-3A-3	1.152±0.008
PUR-2A-4	1.177±0.008	PUR-3A-4	1.135±0.003

Table 5. Density of polyurethanes (results partially presented in [54, 61, 62, 65]).

2.2.6. The oil and water sorption

The implanted material, immersed into a living body is affected by surrounding solutions. Physiological body fluids are constituted of water and floating inorganic and organic compounds, such as lipids. The tendency of water and lipids sorption by polymer is important for its stability in natural conditions. Water plays a key role in the process of hydrolysis whereas lipids accelerate calcification and environmental stress cracking of PUR surfaces. Moreover, the tendency for oil sorption could predispose polymer to albumin sorption. Albumin is peptide absorbable on implant surface what makes the natural junction with natural environment.

The sunflower oil and water sorption by PUR samples were performed at the physiologic temperature of the human body (37°C).

It was stated that the oil sorption by PUR samples based on PTMG and R,S-PHB was much higher than for PURs with PCL and R,S-PHB in soft segments [65, 66].

As it was expected, aliphatic PURs (based on asymmetric diisocyanate) absorbed more oil than aromatic ones. The oil sorption by aliphatic and aromatic PURs were significantly reduced after the introduction of R,S-PHB into soft segments based on PTMG [61, 62]. It could suggest an increase the hydrophility of PURs after R,S-PHB incorporation into their structure. Moreover, it indicated that PURs based on PTMG could be more biocompatible (according to their affinity to lipids) than PURs with PCL.

PUR	Oil sorption ±SD [%]	Water sorption after 24h [%]	Water sorption after 2 weeks [%]	PUR	Oil sorption ±SD [%]	Water sorption after 24h [%]	Water sorption after 2 weeks [%]
PUR-2A-1	4.0±0.16	1.8	3.9	**PUR-3A-1**	6.9±0.28	1.3	1.5
PUR-2A-2	10.1±0.97	1.5	1.3	**PUR-3A-2**	13.4±0.27	2.0	1.5
PUR-2A-3	0.7±0.05	1.1	0.9	**PUR-3A-3**	0.7±0.01	1.1	1.1
PUR-2A-4	1.4±0.32	0.8	0.7	**PUR-3A-4**	0.6±0.22	0.9	1.1

Table 6. The weight changes of PUR samples after incubation in sunflower oil (results partially presented in [61, 62, 65]).

Estimated wetting angle (57°C–71°C) suggested that PURs were hydrophilic [64]. But they absorbed a very low amount of water and only a bit higher in the case of using of PTMG for soft segments building. As mentioned before, the glass temperature of soft segments of polyetherurethanes and polyether-esterurethanes (Table 3) was lower than Tg of polyesterurethanes (PURs with PCL and PCL+R,S-PHB). Lower glass temperature was connected to them being less stiff than polyesterurethanes [67]. In case when the chains were stiff, their mobility was reduced so water could not penetrate easy between them. Higher water absorption by PURs based on PTMG could suggest their higher susceptibility to degradation in aqueous medium. In a PUR network, the particles (such as free radicals and enzymes) that could facilitate its degradation may penetrate with the water.

The presence of R,S-PHB in soft segments increased the water sorption of aromatic PURs [61].

2.2.7. Bacteriostatic properties

The influence of polyurethane PUR-3A-3 on microorganisms (*Staphylococcus aureus, Escherichia coli,* and *Candida albicans*) was investigated using disk methods [68, 69]. The *Staphylococcus aureus* growth around PUR samples was the most inhibited. The bacterial growth was inhibited for 6 mm around a circle sample (diameter of the sample was 8 mm). Knowing that 3-hydroxybutyrate belonging to fatty acids revealed antibacterial activity, it was concluded that R,S-PHB was responsible for the decreasing microorganism growth [68]. But the determined number of survival *Staphylococcus aureus* bacteria directly contacted with PUR sample PUR-3A-3 in a tube method showed only slight decrease of bacteria quantity in comparison to control probe [69]. It suggested its bacteriostatic but no bactericidal properties.

2.2.8. Hemocompatibility

Hemocompatibility of obtained PURs was estimated by observations of changes in morphology and coagulation parameters of whole blood after 4 h of direct contact with polymer samples using flow cytometry and photooptical methods [64, 65].

Parameter [unit]	PUR-2A-3	PUR-3A-1	PUR-3A-3	Blood sample without PUR	Reference value
WBC [*10^3/μl]	6.0	5.9	6.1	6.1	4.0-10.0
RBC [*10^6/ μl]	4.4	4.4	4.4	4.3	4.0-5.0
HGB [g/dl]	12.6	12.7	12.6	12.2	12.0-16.0
HCT [%]	38.1	38.8	37.9	38.0	37.0-47.0
MCV [fl]	87.6	88.0	87.4	87.2	80.0-96.0
MCH [pg]	28.8	28.8	28.8	28.8	27.0-32.0
MCHC [g/dl]	32.9	32.7	33.0	33.0	31.0-36.0
RDW [%]	13.7	13.8	13.9	13.7	11.5-14.5
PLT [*10^3/μl]	255.5	253.4	246.8	247	140.0-400.0
MPV [fl]	9.5	11.3	10.7	9.4	7.0-12.0
APTT [RATIO]	31.3	36.5	29.4	29.8	26.0-37.0
FIBR [g/l]	3.2	1.2	3.5	3.4	1.5-4.5

Table 7. Blood parameters after direct contact with PUR samples and those without contact.

The values of hematologic parameters before and after the incubation of PUR samples in blood were in reference ranges. Differences in comparison to the control probe were not observed that suggests the lack of hemolysis activated by polyurethane presence. Insignificant changes in platelet and fibrinogen concentration and in APPT during direct contact of blood with polymer samples suggested that polyurethanes based on synthetic poly([R,S]-3-hydroxybutyrate) could be atrombogenic [65].

2.2.9. Degradability

The influence of the surrounding environment on implanted material can be simulated using hydrolytic and oxidative solutions. Also, simulated body fluids (SBF) and Ringer solutions included ions that could be found in natural fluids are often used for estimation of biomaterials degradability.

Deionized water or phosphate buffer solution (PBS) is generally used for obtaining hydrolytic conditions in investigations of polymers degradation. According to Christenson et al. [70], the degradation of PURs in a solution of $CoCl_2/H_2O_2$ effectively reflected the oxidation occurring in the living body. The similar changes in the structure of the polymers after one year implantation in the body of rats and after 24-day action of the oxidation mixture were observed. It has been found that the arrangement of $CoCl_2/H_2O_2$ degraded the soft segments of the polyetherurethanes 17 times faster and the polycarbonate urethanes - 14 times [70]. Aliphatic PUR were also degraded in SBF and Ringer solution.

PUR	The weight changes ±SD [%]			
	36 weeks of incubation in PBS	16 weeks of incubation in $CoCl_2/H_2O_2$	36 weeks of incubation in SBF	36 weeks of incubation in Ringer
PUR-2A-1	-1.4±0.1	-22.4±0.9	not estimated	
PUR-2A-2	0.4±0.3	-34.3±0.5		
PUR-2A-3	-3.2±0.1	-30.2±3.3		
PUR-2A-4	0.9±0.2	-7.5±0.7		
PUR-3A-1	-60.3±13.2	-70.9±1.7	disintegration	-16.1±3.5
PUR-3A-2	-55.2±2.3	-29.7±4.3	10.0±2.5	-0.8±0.1
PUR-3A-3	-5.2±0.1	-11.2±0.4	-36.6±1.8	-13.6±0.1
PUR-3A-4	1.4±0.1	-8.2±0.1	6.2±1.0	-1.2±0.3

Table 8. The weight changes of PUR samples after incubation (at 37°C) in phosphate buffer solution (PBS), oxidative solution ($CoCl_2/H_2O_2$), simulated body fluids (SBF), and Ringer solution (Ringer) (results partially presented in [52, 54, 71]).

The susceptibility to hydrolytic and oxidative degradation of obtained PUR with synthetic poly([R,S]-3-hydroxybutyrate) indicated that these materials were more sensitive to the oxidative than hydrolytic conditions. Using an aliphatic diisocyanate in the synthesis (instead of aromatic) increased the susceptibility of PURs to degradation, especially in hydrolytic environments. More susceptible to degradation processes were PUR with PTMG than PCL in soft segment [71].

Introduction of R,S-PHB into the soft segments increased the rate of degradation in all investigated solutions. Aliphatic PURs based on R,S-PHB and PTMG appeared as the most sensitive to conditions of all degradative solutions (higher sample mass loss and molecular weight reduction were noticed) [54].

In some cases, after 36 weeks of incubation the sample mass did not change significantly or even increased. The observed molecular weight reduction after incubation of PURs based on PCL in phosphate buffer indicated that ester linkages were hydrolyzed but because of the insolubility of PCL products, they were not rinsed polymer bulk [54]. Mondal et al. [55] suggested that the degradation products could be retained in bulk films by hydrogen bonding, van der Waals force, polar interaction, etc.

Meanwhile, increasing of sample mass of PUR-3A-2 and PUR-3A-4 after incubation in SBF was probably the result of salts molecules trapping between the macrochains of the polymer network what influenced on the samples mass [52]. Moreover, according to microscope observations of polymer samples presented earlier presence of R,S-PHB in soft segments of PURs protected them against the salt sediments.

Degradability of PURs with R,S-PHB in soft segments could be also controlled by their mixing with PLA [66]. The presence of PLA in polyurethane blends accelerated their degradation in hydrolytic, trypsin, and lipase solutions. The significant reduction of molecular weight of polymer samples after incubation in phosphate buffer and the lack of mass changes after incubation in enzyme solutions suggested that polyurethanes and their blends were degraded via chemical hydrolysis. The investigations of morphology of the surface structure, which was changed after the incubation in both enzymes indicated that the enzymatic hydrolysis had been already initiated [66].

2.2.10. Electrospinning

DSC and WAXS results indicated that similar PURs (with molar ratio of NCO:OH=2:1 in prepolymer) containing PCL in soft segments had higher ability for crystallization than those having PTMG in soft segments [53]. It was the reason why PURs containing PTMG and R,S-PHB in soft segments was chosen for electrospinning. Polymer appeared as spinnable in an electric field, with thermal stability (no phase transitions) in the temperature range up to 95°C. Electrospinning of polyether-esterurethane from hexafluoro-2-propanole solution resulted in the formation of fibers with an average diameter ca. 2 μm. [53].

3. Conclusion

It could be stated that PURs based on synthetic poly([R,S]-3-hydroxybutyrate) displayed the properties appropriate for further investigations for medical applications such as degradable scaffolds. Properties of presented polyester- and polyester-etherurethanes suggest that they could be biocompatible, biostatic, and biodegradable under conditions of living body environment. It suggests also that aromatic diisocyanate may be successfully replaced by an aliphatic one. Incorporation of synthetic polyhydroxybutyrate into soft segments of PURs decreased their degree of crystallinity and increased degradability.

Using polycaprolactonediol for PUR synthesis is appropriate for the design of material undergoing slow and gradual degradation in living body. Higher oil sorption and faster rate of degradation of aliphatic PURs based on polyoxytetramethylenediol promote it for being used as biodegradable scaffolds with hydrophobic active substance.

Author details

Joanna Brzeska

Address all correspondence to: j.brzeska@wpit.am.gdynia.pl

Gdynia Maritime University, Faculty of Entrepreneurship and Quality Science, Department of Chemistry and Industrial Commodity Science, Gdynia, Poland

References

[1] Król P. Polyurethanes-A review of 60 years of their syntheses and applications (in Polish). Polimery. 2009;54:489–500.

[2] Hepburn C. Polyurethane Elastomers. 2nd ed. London, New York: Elseviers Science Publishers Ltd.; 1992. 442 p.

[3] Wirpsza Z. Polyurethanes: Chemistry, technology, and applications (in Polish). Warszawa: WNT; 1991.

[4] Pangman WJ. Compound prosthesis device. US Patent. 1965;US 2.842.775.

[5] Pangman WJ. Compound prosthesis. US Patent. 1965;US 3.189.921.

[6] De la Pena-Salcedo JA, Soto-Miranda MA, Lopez-Salguero JF. Back to the future: A 15-Year experience with polyurethane foam-covered breast implants using the partial-subfascial technique. Aesth Plast Surg. 2012;36:331-338. DOI: 10.1007/s00266-011-9826-5.

[7] Yari A, Yeganeh H, Bakhshi H. Synthesis and evaluation of novel absorptive and antibacterial polyurethane membranes as wound dressing. J Mater Sci: Mater Med. 2012;23:2187–2202. DOI: 10.1007/s10856-012-4683-6.

[8] Filardo G, Zaffagnini S, Di Martino A, Di Matteo B, Muccioli GMM, Busacca M, Marcacci M. Biodegradable polyurethane meniscal scaffold for isolated partial lesions or as combined procedure for knees with multiple comorbidities: clinical results at 2 years. Knee Surg Sport Traumatol Arthrosc. 2014;22:128-134. DOI: 10.1007/s00167-012-2328–4.

[9] Morral-Ruíz G, Melgar-Lesmes P, García ML. Solans C. Garcia-Celma ML. Polyurethane and polyurea nanoparticles based on polyoxyethylene castor oil derivative surfactant suitable for endovascular applications. International Journal of Pharmaceutics. 2014; 461:1–13. DOI: org/10.1016/j.ijpharm.2013.11.026.

[10] Macocinschi D, Filip D, Vlada S, Oprea AM, Gafitanu CA, Characterization of a poly(ether urethane)-based controlled release membrane system for delivery of ketoprofen. Applied Surface Science. 2012;259:416–423. DOI: org/10.1016/j.apsusc.2012.07.060.

[11] Machado HB, Correia RN, Covas JA. Synthesis, extrusion and rheological behaviour of PU/HA composites for biomedical applications. J Mater Sci: Mater Med. 2010;21:2057–2066. DOI: 10.1007/s10856-010-4079-4

[12] Jung YC, Cho JW. Application of shape memory polyurethane in orthodontic. J Mater Sci: Mater Med. 2010;21:2881–2886. DOI: 10.1007/s10856-008-3538-7.

[13] El-Sayed, El Gabry LK, Allam OG. Application of prepared waterborne polyurethane extended with chitosan to impart antibacterial properties to acrylic fabrics. J Mater Sci: Mater Med. 2010;21:507–514. DOI: 10.1007/s10856-009-3900-4.

[14] Khandwekar AP, Doble M. Physicochemical characterisation and biological evaluation of polyvinylpyrrolidone-iodine engineered polyurethane (Tecoflex). J Mater Sci: Mater Med. 2011;22:1231–1246. DOI:10.1007/s10856-011-4285-8.

[15] Park D, Larson AM, Klibanov AM, Wang Y. Antiviral and Antibacterial Polyurethanes of Various Modalities. Appl Biochem Biotechnol. 2013;169:1134–1146. DOI: 10.1007/s12010-012-9999-7.

[16] Xu H, Chang J, Chen Y, Fan H, Shi B. Asymmetric polyurethane membrane with inflammation responsive, antibacterial activity for potential wound dressing application. J Mater Sci. 2013;48:6625–6639. DOI: 10.1007/s10853-013-7461-z.

[17] Akbarian M, Olya ME, Ataeefard M, Mahdavianc M. The influence of nanosilver on thermal and antibacterial properties of a 2 K waterborne polyurethane coating. Progress in Organic Coatings. 2012;75:344–348. DOI.org/10.1016/j.porgcoat.2012.07.017.

[18] Wang J., Sun P., Zheng Z., Wang F. Wang X. Glutathione-responsive biodegradable polyurethanes based on dithiodiundecanol. Polymer Degradation and Stability. 2012;97:2294–2300. DOI.org/10.1016/j.polymdegradstab.2012.07.041.

[19] Da Silva GR, da Silva-Cunha A Jr, Francine Behar-Cohen F, Ayres E, Oréfic RL. Biodegradable polyurethane nanocomposites containing dexamethasone for ocular route. Materials Science and Engineering. 2011;C31:414–422. DOI: 10.1016/j.msec. 2010.10.019.

[20] Saralegi A, Gonzalez ML, Valea A, Eceiza A, Corcuera MA. The role of cellulose nanocrystals in the improvement of the shape-memory properties of castor oil-based segmented thermoplastic polyurethanes. Composites Science and Technology. 2014; 92:27–33. DOI.org/10.1016/j.compscitech.2013.12.001.

[21] Chan-Chan LH, Tkaczyk C, Vargas-Coronado RF, Cervantes-Uc JM, Tabrizian M, Cauich-Rodriguez JV. Characterization and biocompatibility studies of new degradable poly(urea)urethanes prepared with arginine, glycine or aspartic acid as chain extenders. J Mater Sci: Mater Med. 2013;24:1733–1744. DOI: 10.1007/s10856-013-4931-4.

[22] Jia L, Prabhakaran MP, Qin X, Kai D, Ramakrishna S. Biocompatibility evaluation of protein-incorporated electrospun polyurethane-based scaffolds with smooth muscle cells for vascular tissue engineering. J Mater Sci. 2013;48:5113–5124. DOI: 10.1007/ s10853-013-7359-9.

[23] Cyras VP, Soledad CM, Analı´ V. Biocomposites based on renewable resource: Acetylated and non acetylated cellulose cardboard coated with polyhydroxybutyrate. Polymer. 2009;50:6274–6280. DOI: 10.1016/j.polymer.2009.10.065.

[24] Foster LJR, Tighe BJ. Centrifugally spun polyhydroxybutyrate fibres: Accelerated hydrolytic degradation studies, Polymer Degradation and Stability. 2005;87:1–10 DOI: 10.1016/j.polymdegradstab.2003.11.012.

[25] Defoirdt T, Boon N, Sorgeloos P, Verstraete W, Bossier P. Short-chain fatty acids and poly-ß-hydroxyalkanoates: (New) Biocontrol agents for a sustainable animal production, Biotechnology Advances. 2009;27:680–685. DOI: 10.1016/j.biotechadv.2009.04.026.

[26] Ali AQ, Kannan TP, Azlina Ahmad, Samsudin R. In vitro genotoxicity tests for polyhydroxybutyrate – A synthetic biomaterial. Toxicology in Vitro. 2008;22:57–67. DOI: 10.1016/j.tiv.2007.08.001.

[27] Lee J, Jung SG, CS, Kim HY, Batt CA, Kim YR. Tumor-specific hybrid polyhydroxybutyrate nanoparticle: Surface modification of nanoparticle by enzymatically synthesized functional block copolymer. Bioorganic & Medicinal Chemistry Letters. 2011;21:2941–2944. DOI: 10.1016/j.bmcl.2011.03.058

[28] Medvecký L, Štulajterová R, Kutsev SV. Microstructure and properties of polyhydroxybutyrate–calcium phosphate cement composites. Chemical Papers. 2011;65 (5): 667–675. DOI: 10.2478/s11696-011-0044-z.

[29] Medvecky L, Giretova M, Stulajterova R. Properties and in vitro characterization of polyhydroxybutyrate–chitosan scaffolds prepared by modified precipitation method. J Mater Sci: Mater Med. 2014;25:777–789. DOI: 10.1007/s10856-013-5105-0.

[30] Tai HY, Fu E, Cheng LP, Don TM. Fabrication of asymmetric membranes from polyhydroxybutyrate and biphasic calcium phosphate/chitosan for guided bone regeneration. J Polym Res. 2014;21:421. DOI: 10.1007/s10965-014-0421-8.

[31] Liu Y, Wang M. Developing a composite material for bone tissue repair. Current Applied Physics. 2007;7:547–554. DOI: 10.1016/j.cap.2006.11.002.

[32] Unverdorben M, Spielberge A, Schywalsky M, Labahn D, Hartwig S, Schneider M, et al. Polyhydroxybutyrate Biodegradable Stent: Preliminary Experience in the Rabbit. Cardiovasc Intervent Radiol, 2002;25:127–132. DOI: 10.1007/s00270-001-0118-3.

[33] Shishatskaya EI, Voinova ON, Goreva AV, Mogilnaya OA, Volova TG. Tissue Reaction to Intramuscular Injection of Resorbable Polymer Microparticles. Bulletin of Experimental Biology and Medicine. 2007;144(6) BIOPHYSICS AND BIOCHEMISTRY.

[34] Shishatskaya EI, Voinova ON, Goreva AV, Mogilnaya OA, Volova TG. Biocompatibility of polyhydroxybutyrate microspheres: in vitro and in vivo evaluation. J Mater Sci: Mater Med. 2008;19:2493–2502. DOI: 10.1007/s10856-007-3345-6.

[35] Althuri A, Mathew J, Sindhu R, Banerjee R, Pandey A, Binod P. Microbial synthesis of poly-3-hydroxybutyrate and its application as targeted drug delivery vehicle. Bioresource Technology. 2013;145:290–296. DOI.org/10.1016/j.biortech.2013.01.106.

[36] Hong SG, Hsu HW, Ye MT. Thermal properties and applications of low molecular weight polyhydroxybutyrate. J Therm Anal Calorim. 2013;111:1243–1250. DOI: 10.1007/s10973-012-2503-3.

[37] Hong SG, Gau TK, Huang SC. Enhancement of the crystallization and thermal stability of polyhydroxybutyrate by polymeric additives. J Therm Anal Calorim. 2011;103:967–975. DOI 10.1007/s10973-010-1180-3.

[38] Arslan H, Adamus G, Hazer B, Kowalczuk M. electrospray ionisation tandem mass spectrometry of poly[(R,S)-3-hydroxybutanoic acid] telechelics containing primary hydroxy end groups. Rapid Commun. Mass Spectrom. 1999;13:2433–2438.

[39] Jedliński Z, Kurcok P, Kowalczuk M. Method of synthesis of amorphous poly([R,S]-3-hydroxybutyric acid), Polish Pat Appl P-339795, 2000.

[40] Piddubnyak V, Kurcok P, Matuszowicz A, Głowala M, Fiszer-Kierzkowska A, Jedliński Z et al. Oligo-3-hydroxybutyrates as potential carriers for drug delivery. Biomaterials. 2004;25:5271–5279. DOI: 10.1016/j.biomaterials.2003.12.029.

[41] Brzeska J, Dacko P, Janeczek H, Janik H, Rutkowska M, Sikorska W, et al. Synthesis, properties and applications of new (bio)degradable polyester urethanes (in Polish). Polimery. 2014;59(5):363–371.

[42] Kunze C, Bernd HE, Androsch R, Nischan C, Freier T, Kramer S, et al. In vitro and in vivo studies on blends of isotactic and atactic poly(3-hydroxybutyrate) for development of dura substitute material. Biomaterials. 2006;27:192–201. DOI: 10.1016/j.biomaterials.2005.05.095.

[43] Freier T, Kunze C, Nischan C, Kramer S, Sternberg K, Saß M, et al. In vitro and in vivo degradation studies for development of biodegradable patch based on poly(3-hydroxybutyrate). Biomaterials. 2002;23:2649–2657.

[44] Cebulska A, Jedliński Z. Preparation of spherical particles of 3-hydroxybutyric acid oligomer (in Polish). Polimery. 2006;6:436–441.

[45] Aziz MSA, Naguib HF, Saad GR. Non-isothermal crystallization kinetics of bacterial poly(3-hydroxybutyrate) in poly(3-hydroxybutyrate-co-butylene adipate) urethanes. Thermochimica Acta. 2014;591:130–139. dx.doi.org/10.1016/j.tca.2014.07.026 0040-6031..

[46] Pan J, Li G, Chen Z, Chen X, Zhu W, Xu K. Alternative block polyurethanes based on poly(3-hydroxybutyrate-co-4-hydroxybutyrate) and poly(ethylene glycol). Biomaterials. 2009;30:2975–2984. doi:10.1016/j.biomaterials.2009.02.005.

[47] Ou W, Qiu H, Chen Z, Xu K. Biodegradable block poly(ester-urethane)s based on poly(3-hydroxybutyrateco-4-hydroxybutyrate) copolymers. Biomaterials. 2011;32:3178–3188. DOI: 10.1016/j.biomaterials.2011.01.031.

[48] Loh XJ, Tan KK, Li X, Li J. The in vitro hydrolysis of poly(ester urethane)s consisting of poly[(R)-3-hydroxybutyrate] and poly(ethylene glycol). Biomaterials. 2006;27:1841–1850. DOI: 10.1016/j.biomaterials.2005.10.038.

[49] Hong JH, Jeon HJ, Yoo JH, Yu WR, Youk JH. Synthesis and characterization of biodegradable poly(3-caprolactone-co-b-butyrolactone)based polyurethane. Polymer Degradation and Stability. 2007;92:1186–1192. DOI: 10.1016/j.polymdegradstab.2007.04.007.

[50] Brzeska J, Dacko P, Janik H, Kowalczuk M, Rutkowska M, Biodegradable polyurethanes and the way of their obtaining. Polish Patent No. 212763 (B1). 2012.

[51] Brzeska J, Dacko P, Gębarowska K, Janik H, Kaczmarczyk B, Kasperczyk J, Kowalczuk M, Rutkowska M. The structure of novel polyurethanes containing synthetic poly[(R,S)-3-hydroxybutyrate], Journal of Applied Polymer Science. 2012;125(6): 4285–4291. DOI: 10.1002/app.36599.

[52] Brzeska J, Heimowska A, Janeczek H, Kowalczuk M, Rutkowska M. Polyurethanes based on atactic poly[(R,S)-3-hydroxybutyrate]: preliminary degradation studies in simulated body fluids. Journal of Polymers and the Environment. 2014;22(2):176–182. DOI: 10.1007/s10924-014-0650-2.

[53] Sajkiewicz P, Brzeska J, Denis P, Sikorska W, Kowalczuk M, Rutkowska M. The preliminary studies of a structure and Electrospinning of new polyurethanes based on synthetic atactic poly[(R,S)-3-hydroxybutyrate]. Bulletin of The Polish Academy of Sciences Technical Sciences. 2014; 62(1):55–60, DOI: 10.2478/Bpasts-2014-0006.

[54] Brzeska J, Janeczek H, Janik H, Kowalczuk M, Rutkowska M. Degradability in vitro of polyurethanes based on synthetic atactic poly[(R,S)-3-hydroxybutyrate]. Bio-Medical Materials and Engineering 25 (2015) 117–125, DOI 10.3233/BME-151262

[55] Mondal S, Martin D. Hydrolytic degradation of segmented polyurethane copolymers for biomedical applications. Polymer Degradation and Stability. 2012;97:1553–1561. DOI: 10.1016/j.polymdegradstab.2012.04.008.

[56] Fei B, Chen Ch, Wu H., Peng Sh., Wang X, Dong L. Quantitative FTIR study of PHBV/bisphenol A blends. European Polymer Journal. 2003;39:1939–1946.

[57] Zhang J, Sato H, Tsuji H, Noda I, Ozaki Y. Differences in the CH_3 O=C interactions among poly(L-lactide). poly(L-lactide)/poly(D-lactide) stereocomplex, and poly(3-hydroxybutyrate) studied by infrared spectroscopy. Journal of Molecular Structure. 2005;735-736:249–257. DOI: 10.1016/j.molstruc.2004.11.033.

[58] Wu Q, Tian G.S, Sun SQ, Noda I, Cheb GQ. Study of microbial polyhydroxyalkanoates using two-dimensional Fourier-transform infrared correlation spectroscopy. J. Appl. Polym. Sci. 2001;82:934–941.

[59] Bayarı S, Severcan F. FTIR study of biodegradable biopolymers: P(3HB), P(3HB-co4HB) and P(3HB-co-3HV). Journal of Molecular Structure. 2005;744-747:529–534. DOI:10.1016/j.molstruc.2004.12.029.

[60] Yilgor I, Yilgor E, Guler IG, Ward TC, Wilkes GL. FTIR investigation of the influence of diisocyanate symmetry on the morphology development in model segmented polyurethanes. Polymer. 2006;47:4105–4114. DOI: 10.1016/j.polymer.2006.02.027.

[61] Brzeska J, Dacko P, Janeczek H, Kowalczuk M, Janik H, Rutkowska M. The influence of synthetic polyhydroxybutyrate on selected properties of novel polyurethanes for medical applications. Part I. Polyurethanes with aromatic diisocyanates in hard segments (in Polish). Polimery. 2010;1:44–47.

[62] Brzeska J, Dacko P, Janeczek H, Kowalczuk M, Janik H, Rutkowska M. The influence of synthetic polyhydroxybutyrate on selected properties of novel polyurethanes for medicine B. Polyurethanes with aliphatic diisocyanate in hard segment (in Polish). Polimery. 2011;1:27–34.

[63] Janik H. Supermolecular structure and selected properties of branched and cross-linked poly(esterurethanes), poly(etherurethanes) and poly(ureauretanes) formed reactively (in Polish). Zeszyty Naukowe Politechniki Gdańskiej. 2005. 599. Wydawnictwo Politechniki Gdańskiej. Gdańsk.

[64] Brzeska J. The influence of synthetic polyhydroxybutyrate on properties of new polyurethanes for medical application [dissertation] (in Polish). Gdynia Maritime University:2010. 189

[65] Brzeska J, Janik H, Kowalczuk M, Rutkowska M. Preliminary investigations of biocompatibility of polyurethanes based on synthetic polyhydroxybutyrate, Engineering of Biomaterials. 2011;106-108(XIV):65–72.

[66] Brzeska J, Heimowska A, Sikorska W, Jasinska-Walc L, Kowalczuk M, Rutkowska M. Degradation of polyurethane/polylactide blends. International Journal of Polymer Science. Forthcoming.

[67] Jonquières A, Clément R, Lochon P. Permeability of block copolymers to vapours and liquids, Progress in Polymer Science. 2002;27:1803–1877.

[68] Brzeska J, Janik H, Kowalczuk M, Rutkowska M. Influence of polyurethanes based on synthetic poly([R,S]-3-hydroxybutyrate) on microorganisms growth. Engineering of Biomaterials. 2011;106-108(XIV):73–78.

[69] Brzeska J, Kowalczuk M, Janik H, Rutkowska M. The properties of polyurethanes based on synthetic polyhydroxybutyrate for medical application. Joint Proceedings Gdynia Maritime University-Bremerhaven. 2012;74:5–14.

[70] Christenson EM, Patel S, Anderson JM, Hiltner A. Enzymatic degradation of poly(ether urethane) and poly(carbonate urethane) by cholesterol esterase. Biomaterials. 2006;27:3920–3926. DOI:10.1016/j.biomaterials.2006.03.012.

[71] Brzeska J, Dacko P, Heimowska A, Janik H, Kowalczuk M, Rutkowska M. Degradability of polyurethanes with synthetic polyhydroxybutyrate in oxidative and hydrolytic environments (in Polish), Ochrona przed korozją. 2012;1:8–14.

Chapter 2

Synthesis and Properties of Multiblock Terpoly(Ester-Aliphatic-Amide) and Terpoly(Ester-Ether-Amide) Thermoplastic Elastomers with Various Chemical Compositions of Ester Block

Joanna Rokicka and Ryszard Ukielski

Additional information is available at the end of the chapter

http://dx.doi.org/10.5772/61215

Abstract

Two series of thermoplastic elastomers with various chemical compositions of ester block were prepared via the reaction of α,ω-dicarboxylic oligo(laurolactam) (PA12, Mw≈2000 g/mol) with oligo(oxytetramethylene)diol (PTMO, Mw≈1000 g/mol) or linoleic alcohol dimer (DLAol) and with dimethyl terephthalate and a low molecular weight glycol (forming during the synthesis of the ester block). The degree of polycondensation (DP_{GT}) of poly(multi-methylene terephtalate) equals to DP_{GT}=2. The influence of the number of carbons separating the terephtalate groups, as well as the effect of *meta-* or *para-* positions of the ester groups in the benzene ring of other blocks, on the synthesis, properties and structure of these elastomers have been evaluated. A nuclear magnetic resonance spectroscopy to carbon (^{13}C NMR) and Fourier transform infrared spectroscopy (FT-IR) were used to confirm their assumed chemical structure. The influence of chemical compositions of ester block on the functional properties and on the values of phase transition temperatures of the products have been determined. The thermal properties and the phase separation of obtained systems were defined by differential scanning calorimetry (DSC), dynamic mechanical thermal analysis (DMTA), wide-angle x-ray diffraction (WAXS) and other standard physical methods. The mechanical and elastic properties of obtained polymers were evaluated.

Keywords: Poly(ester-ether-amide), poly(ester-aliphatic-amide), multiblock terpolymers, elastomers, phase structure

1. Introduction

A growing demand for polymeric materials in the packaging, sport, automotive, and medicine industries stimulates the search for innovative materials with thermal and mechanical properties individually tailored for a given field. Depending on the application, these materials have to be characterized, among others, by their resistance to chemical, mechanical, or biological factors. These requirements may be successfully satisfied by the thermoplastic multiblock elastomer (TPE).

They combine the end-use physical properties of vulcanized rubbers with the easy processing of thermoplastic [1–5]. The properties of TPEs are influenced by an appropriate phase structure and its thermal reproducibility in the heating-cooling cycle, functional qualities (large, reversible deformations) and the processing properties (possibility of multiple melting and solidifications). TPEs are considered as polymeric materials, in which, as a result of the phase separation, at least two phases (soft and hard) are distinguished, which must be thermodynamically immiscible to prevent the interpenetration. Hence, these are plastics possessing at least the two values of the physical transition temperatures, i.e., glass transition temperature T_g and melting point temperature T_m or T_{g1} and T_{g2}. These two temperatures determine the points at which the particular elastomer goes through transitions in its physical properties [6]. There are three distinct regions:

- at very low temperatures, below the T_g of the soft phase, in which both phases are hard, so the material is stiff and brittle;
- between the T_g of the soft phase and the T_m of the hard phase, in which material is soft and elastic, resembling a conventional vulcanized rubber. In these temperature range a modulus of elasticity stays relatively constant and this region is referred as "rubbery plateau" which is characteristic for TPEs;
- above the T_m of the hard phase, where the hard phase softens and melts and the material becomes a viscous fluid [7, 8].

Some of multiblock copolymers that have a heterophase internal structure are classified as the group of thermoplastic elastomers. Their macromolecule must contain two types of chain segments: amorphous, referred to as soft blocks, and hard blocks, which are mostly crystalline [9–12]. These blocks differ considerably in their physical and chemical properties. The soft blocks, which are capable of forming a soft-phase matrix, provides the elastomeric character, susceptibility to hydrolysis, and behavior of copolymer at low temperatures; while the hard segments determine processability, mechanical properties, hardness, and high temperature resistance. Hard segments are able to form intermolecular association with other hard blocks and these blocks form the domains of hard phase and are immersed in a soft-phase matrix. To classify the block elastomers to TPEs, their internal structure must comply with strict conditions. The soft phase should exhibit a relatively small elastic modulus, relatively low glass transition temperature, and a lower density. Moreover, these blocks should ensure weak intermolecular interactions and a large capability for motion and rotation of short sequences of chains. The hard phase must possess a relatively large elasticity modulus, high glass transition temperature or melting point, and high density. The segments that build this phase must exhibit a tendency to aggregate with the same kind of segments. This aggregation leads to the thermally reversible "physical cross-linking", which is stabilized by the van der Waals forces, high cohesive energy, hydrogen bonds, ionic bonds, the polar and dispersive interactions, or the ability to crystallization. The intermolecular interactions of rigid blocks affect the stabilization of the phase structure of the whole polymeric system. The hard segments must have a larger cohesive energy than the flexible segments and hence, a higher thermodynamic potential. The potential difference is a force that induces and stabilizes a heterophase structure [13–17].

The block copolymer will exhibit characteristics of a good elastomer if it complies with five inseparable conditions:

- macromolecules must be linear or weakly branched, which is responsible for thermoplasticity of these copolymer. Such macromolecules construction allows a close proximity of the hard blocks and intermolecular interactions;
- chemical composition of the blocks must have well-defined differences such that the thermodynamic condition of phase separation was fulfilled and the temperature spectrum of Young's modulus had a wide plateau of elasticity;
- it must have appropriate phase composition depending on the type of blocks and their dimensions, which is responsible for the high elasticity features;
- it must possess an appropriate phase dispersion;
- it must have an appropriate interphase density, which is responsible for stabilization of the nanostructure.

The TPE properties are a result of the combination of the individual features of the respective blocks, hence a change of their chemical structure or their relative mass fraction, enables a modification of the macromolecule properties in the desired directions [18–25].

In the present chapter, the synthesis of multiblock thermoplastic elastomers and the relationship between the chemical structure of soft block and the properties in connection with the phase structure of terpolymers are described. The following terpolymers were selected for this research study:

- poly[(multi-methylene terephthalate)-block-(oxytetramethylene)-block-(laurolactam)] xGT-PTMO-PA12
- poly[(multi-methylene terephthalate)-block-(linoleic alcohol dimer)-block-(laurolactam)] xGT-DLAol-PA12

2. Experimental part

2.1. Materials

The following substrates were used: dimethyl terephthalate (DMT) - Chemical Plant "Elana", ethylene glycol (2G), 1,3-propanediol (3G), 1,4-butanediol (4G), 1,5-penatnediol (5G), 1,6-

hexanediol (6G) - Sigma-Aldrich, poly(oxytetramethylene)diol with molecular weight 1000 g/mol (PTMO) - Du Pont, linoleic alcohol dimer (DLAol) - Croda, a titanate catalyst (TiO_2/SiO_2) - Sachtleben Chemic GmgH, thermal stabilizer (Irganox 1010) - Ciba Geigy, dodecano-12-lactam, sebacic acid - Aldrich Chemie. The lactam and the dicarboxilic acid are the substrates prepared in our laboratory: α,ω-dicarboxilic oligo(laurolactam) (PA12) with the number average molecular weight 2000 g/mol [26, 27].

2.2. Synthesis of oligoamide blocks

The synthesis of oligoamides with molecular weight 2000 g/mol bi-ended with carboxylic groups was carried out.

Synthesis was carried out in a cylindrical shape with conical bottom 6-dm^3 autoclave made of stainless steel. The ratio of height to reactor diameter is h:d=3.5. The heating system comprise a set of three resistance heaters enabling control of temperature in the range 20°C–400°C. The regulators are controlled by Fe-constantan thermocouples and Pt-100 thermoresistors.

A suitable amount of laurolactam and sebacic acid, which was a stabilizator of molecular weight of the obtained oligoamides, was placed in the reactor. A small dose of water and phosphoric acid was added to simplify the initiation of the reaction. Before the synthesis, the autoclave was purged three times with nitrogen at the pressure of 0.5 MPa to gain an oxygen-free reaction environment. After that, the reaction mixture was kept in the reactor under a pressure of 0.1 MPa and then heated to a temperature of 300°C. A pressure would not exceed 1.6 MPa. The pressure was controlled by the removal of water vapor that contained a small amount of amide di- and trimers. This pressure stage of the reaction lasted 5 h. In the last half hour of this stage, the temperature raised to 305°C. Then, the pressure was reduced during 0.5 h to atmospheric pressure without decreasing the temperature. The pressureless stage (polycondensation) was carried out for 5 h under a flow of nitrogen. After the reaction was completed, the obtained oligoamide was extruded by compressed nitrogen into a tube with water that was vigorously stirring with bubbling air. To extract the residual by-products and unreacted lactam the finished product was rinsed three times with boiling water and then with distilled water. After drying in air and in a vacuum dryer at the temperature 60°C, the oligoamide was grounded and characterized.

2.3. Synthesis of multiblock thermoplastic elastomers

Synthesis of block terpolymers proceeded in the two stages. The first stage was the transesterification reaction of dimethyl terephthalete with glycol leading to the formation of polyester and the release of methanol and the esterification (in a separate reactor) of α,ω-dicarboxylic oligo(laurolactam) with oligo(oxytetramethylene)diol or linoleic alcohol dimer (by-product is water) in the presence of catalyst. From the respective amounts of methanol and water, it was concluded that the conversation in the transesterification reaction was 95% and the degree of esterification was 90% (degrees was expressed as the weight ratios of the released methanol or water to the respective stoichiometric amounts of these products). The second stage of the process comprises the specific condensation polymerization of mixed intermediates obtained in the first stage of synthesis. The course and parameters of synthesis is presented in Figure 1.

http://dx.doi.org/10.5772/61215

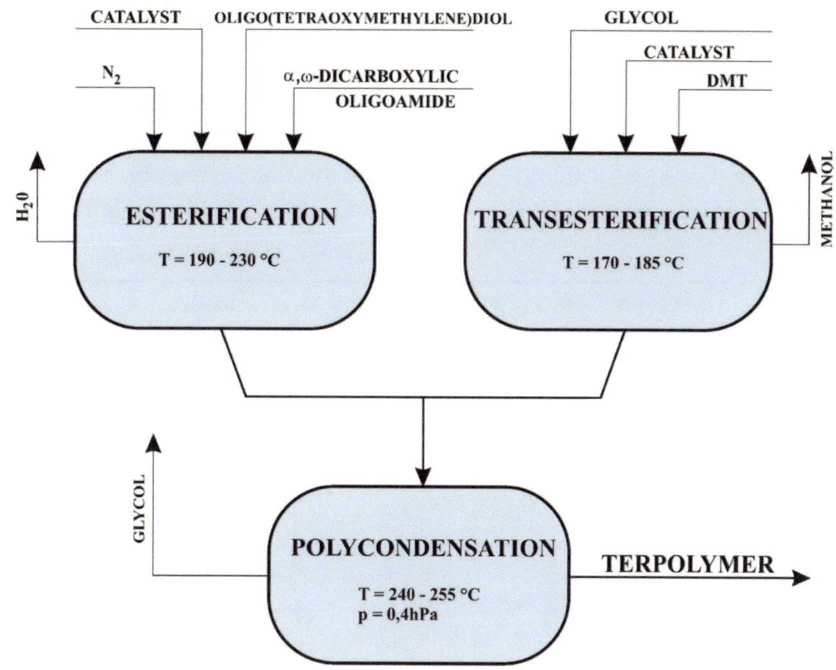

Figure 1. Structure flow chart.

Preparation of terpolymers relies on the replacement in the poly(multi-methylene terephthalate) macromolecule (xGT), a certain part of fragments originated from the terephthalic acid by the dicarboxylic oligoamide block, which was derived from glycol by the oligoetherdiol or alifatic block (Figure 2).

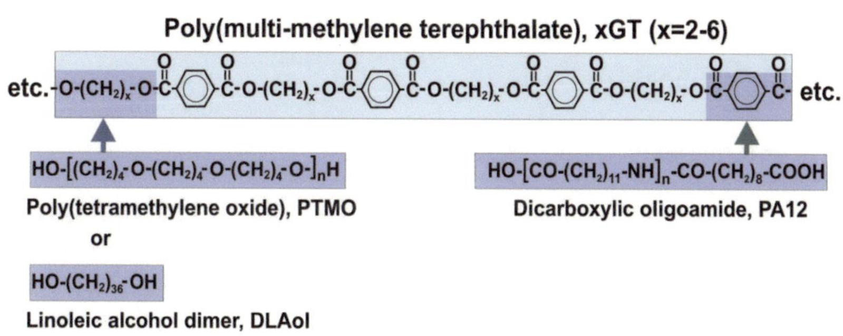

Figure 2. Multi-methylene terephthalate modification.

The previous investigations carried out by Ukielski [28–30] have demonstrated that the molar ratio of PTMO with molecular weight 1000 g/mole to PA12 with molecular weight 2000 g/mole should be ranged between 2 and 3.5. This determines a relatively small fraction of the xGT sequence in the soft phase, large degrees of separation of both soft and hard phase and comparable fraction of the respective blocks in the interphase. Based on the results of previous works, the molar ratio of PTMO/PA12=3 was assumed for further studies. A two series of terpolymers composed of the blocks PTMO or DLAol and PA12 with the constant molar weight amounting to respectively, 1000 g/mole, 570 g/mole and 2000 g/mol and xGT block, with constant degree of polycondensation amounting DP_{xGT}=2, which is formed during the synthesis, were prepared. The series differ in a chemical structure of the ester block. The molar ratio and weight ratios of reagent used for the synthesis were presented in Table 1.

Sample	Glycol	Soft block	m_{soft}	m_{PA12}	m_{DMT}	m_{xG}	w_{PTMO}	w_{xGT}	w_{PA12}	Series
1	2G	PTMO	3	1	5	9	0,520	0,133	0,347	I
2	3G	PTMO	3	1	5	9	0,515	0,141	0,343	
3	4G	PTMO	3	1	5	9	0,510	0,150	0,340	
4	5G	PTMO	3	1	5	9	0,505	0,158	0,337	
5	6G	PTMO	3	1	5	9	0,501	0,166	0,334	
6	2G	DLAol	3	1	5	9	0,382	0,172	0,447	II
7	3G	DLAol	3	1	5	9	0,377	0,182	0,441	
8	4G	DLAol	3	1	5	9	0,373	0,192	0,436	
9	5G	DLAol	3	1	5	9	0,368	0,201	0,430	
10	6G	DLAol	3	1	5	9	0,364	0,211	0,425	

m – molar ratio, w – weight fraction

Table 1. The composition of terpolymers with constant molecular weights of PA12=2000 g/mole and PTMO=1000 g/mole; DP_{xGT}=2.

3. Results and discussion

3.1. Properties of α,ω-dicarboxylic oligoamides

Assumed molecular weight of the oligoamide was 2000 g/mole. The difference between the molecular weight values, calculated from the amounts of the carboxylic groups and the assumed theoretical values do not exceed 2.5% of the experimental error, which supports the correctness of the experimental assumptions. The melting temperature values of the PA12 oligoamides are lowers than the melting temperature of the polyamide 12, which is 179°C.

Usage of sebacic acid as the molecular weight stabilizer leads to dicarboxylic oligoamids. Obtained oligoamids had coherent with assumed molecular weight values, didn't contain amid groups, and may be used as hard block in various types of thermoplastic multiblock elastomers. Some properties of the obtained α,ω-dicarboxilic oligoamides are presented in Table 2.

Sample	[-COOH], μeq/g	[-NH$_2$], μeq/g	M$_n$, g/mole	T$_m$, °C
OA1	1013	4	1974	161-170
OA2	1004	2	1992	162-173
OA3	992	0	2016	167-173
OA4	1032	0	1938	157-167
OA5	1000	2	2000	160-171

[-COOH] – the concentration of carboxyl groups, [-NH$_2$] – the concentration of amino groups, M_n - average molecular weight, T_m – melting temperature range determined by means of the Boethius microscope

Table 2. Properties of obtained oligoamides.

3.2. Properties of TPE

The number of carbons x separating the terephthalate groups in the ester block of TPEs influences all their properties, which were presented in Table 3.

Sample	[η], dL/g	H, Shore A	H, Shore D	p_{H2O}, %	p_{benzen},%	σ, MPa	E, MPa	ε, %	$T_m{}^i$, °C	$T_m{}^f$, °C
1	1,25	84	29	2	134	6,9	18,1	112	135	150
2	1,40	76	26	2	174	4,7	13,6	200	122	135
3	1,68	76	24	0	182	12,5	29	165	157	151
4	1,29	78	24	1	174	6,5	21,1	253	128	135
5	1,32	75	23	3	204	12,1	27	132	146	150
6	1,29	84	19	1	84	13,7	25,4	99	133	148
7	1,55	87	19	1	82	14,5	23,6	81	125	138
8	1,74	87	19	1	75	17,5	46,6	123	141	164
9	1,35	69	16	0	96	13,2	29,8	127	129	150
10	1,25	69	16	4	117	17,4	48,9	104	150	161

[η] – limiting viscosity number, H – hardness, p_{H2O} and $p_{benzene}$ – absorbability of water and benzene, respectively, σ – tensile strength, E – the Young module, ε – elongation at break, $T_m{}^i$,$T_m{}^e$ - initial and end of melting point

Table 3. The properties of terpolymer elastomers TPE with variable chemical structure of ester block.

TPEs have the satisfactory molecular weights and good mechanical properties when the values of their limiting viscosity numbers are larger than [η]>1.25 dL/g. The [η] values of obtained polymers have proved that they are composed of the macromolecules with satisfactory molecular weights. The ability to swell depends on the polarity of the solvent and the chemical structure of the polymer macromolecule. Swelling occurs only in the amorphous structural zones of the terpolymer. It is therefore a measure of the quality and content of the continuous phase. As a physical phenomenon indicates an internal cross-linking of the polymer and describes its physical structure. Absorbability increases with increasing mobility of the macromolecules and decreases with an increase in cohesive energy between them. The obtained results of swelling in water indicate for a hydrophobic character of all the prepared polymers. The swelling does not exceed 4%, which indicates the water penetration in the amorphous phase of polymer only in a slight degree. An increase in the number of carbons separating the terephthalate groups in the ester block of TPEEAs causes an increase of absorbability of benzene and decrease of hardness, due to an increase in amorphous phase content. Flexible blocks create more and better polymer matrix and the content of crystalline phase decreases. It results in the relaxation of the structure and distance from other polymer macromolecules, thus easier benzene penetration into the material and less rigidity.

3.3. Chemical structure of TPEs

The FT-IR spectra [31, 32] of terpolymers selected from each series are shown in Figure 3 and 4. Obtained copolymers had all characteristic bands for esters, aliphates or ether, and amides, which are presented in Table 4.

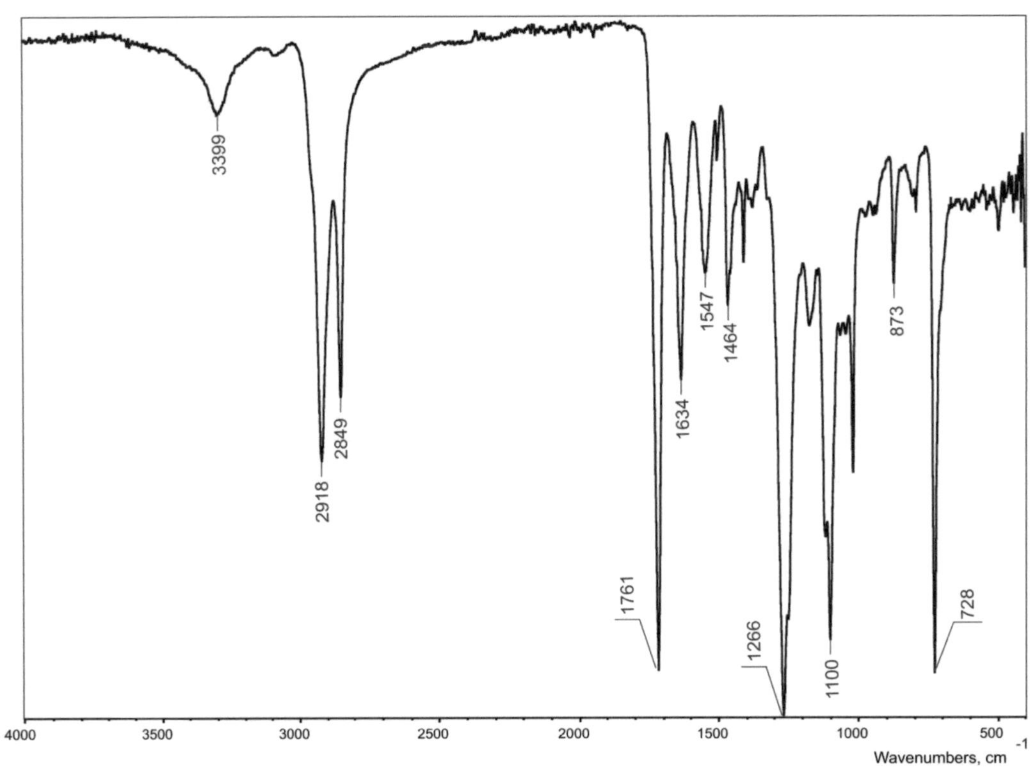

Figure 3. FT-IR spectra for terpolymer 3GT-DLAol-PA12.

http://dx.doi.org/10.5772/61215

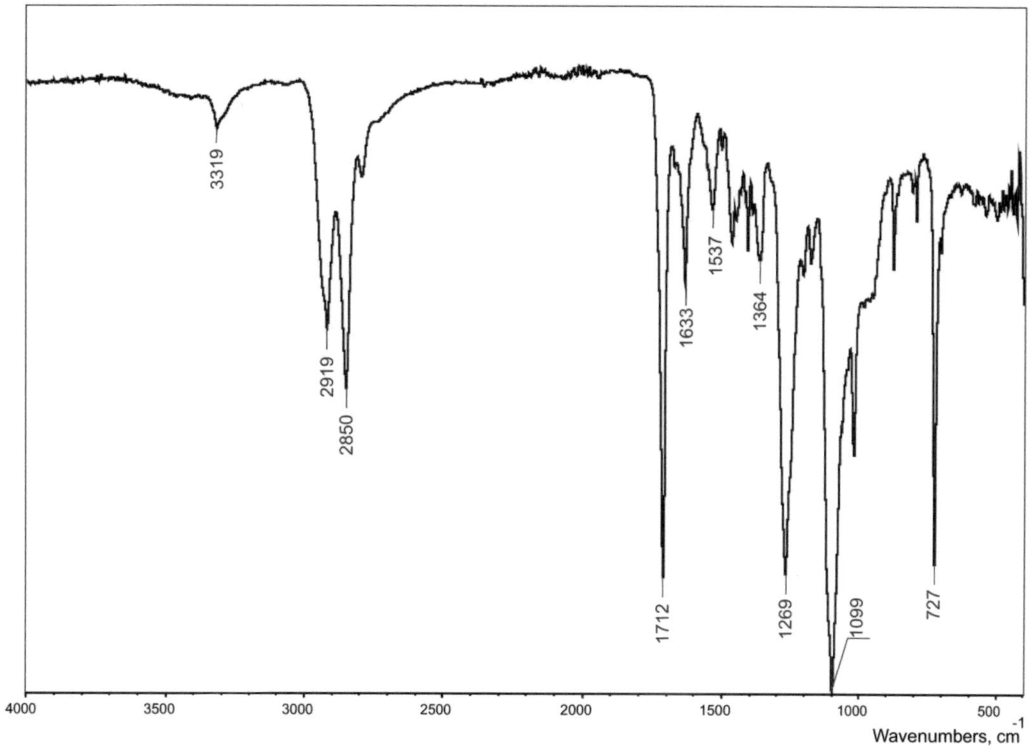

Figure 4. FT-IR spectra for terpolymer 3GT-PTMO-PA12.

3GT-PTMO-PA12 band frequency, ppm	3GT-DLAol-PA12 band frequency, ppm	Chemical structure
3319	3399	Hydrogen-bonded N-H stretching
2919 and 2850	2918 and 2849	CH_2 asymmetric and symmetric stretching
1712	1761	Ester C=O stretching
1633	1634	Amide I, amide C=O stretching
1537	1547	Amide II, C-N stretching + amide C=O in plane bonding
1364	1371	CH bond, CH_2 twisting
1269	1266	Amide III, C-N stretching + amide C=O in plane bonding
1099	1100	C-O-C asymmetric stretching
727	728	Aromatic C-H out of plane bonds

Table 4. Wavenumbers and assignments of FT-IR band exhibited by obtained terpolymers.

Chemical structure of new materials was also verified with ^{13}C NMR spectroscopy [31–34] and the results with peak assignments are presented in Figures 5 and 6. Analysis of the ^{13}C NMR spectra showed the presence of all characteristic groups present in the esters, ethers or fatty acids, and amides. Signals are noted in the range 26.89–218.53 ppm and their interpretation are detailed in Table 5.

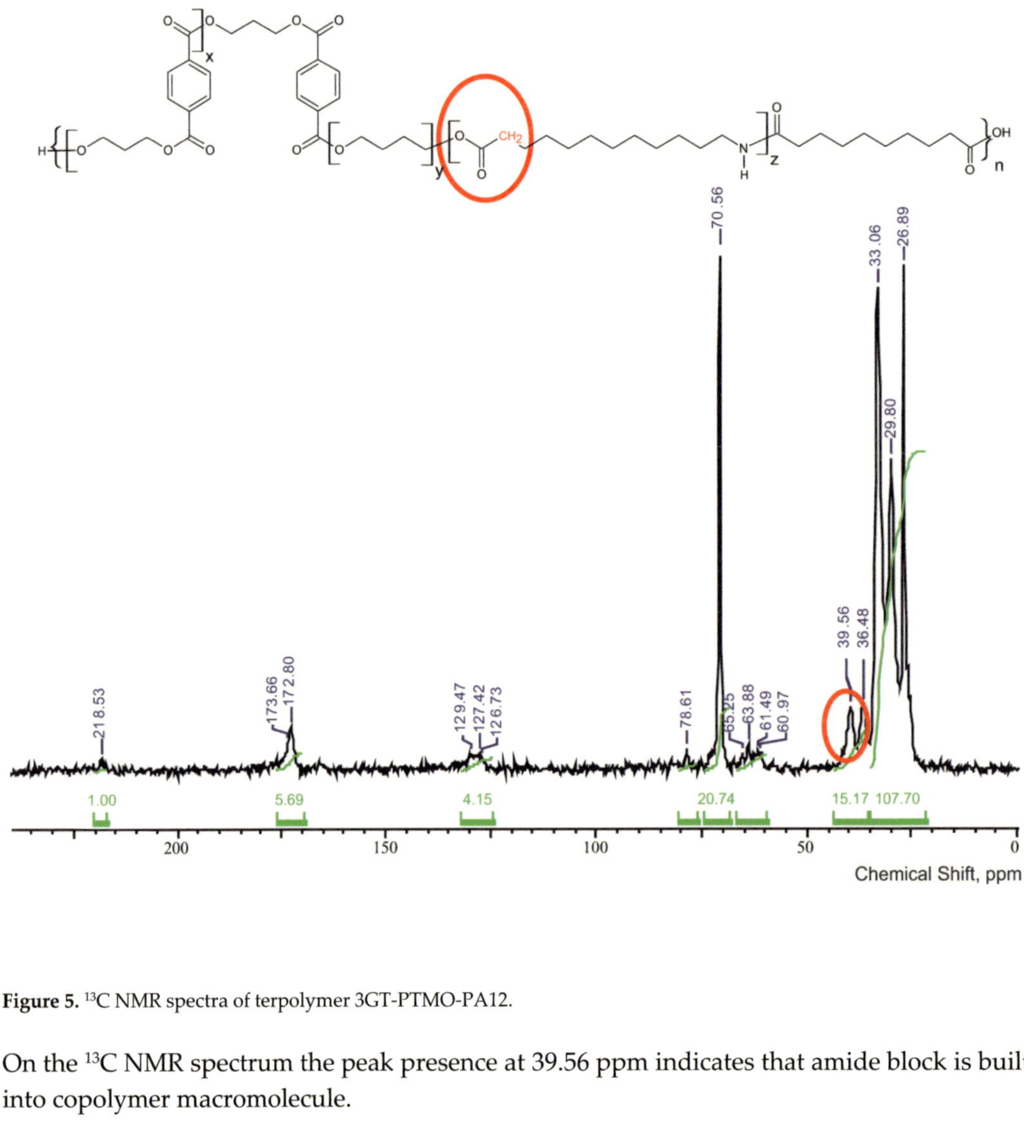

Figure 5. ^{13}C NMR spectra of terpolymer 3GT-PTMO-PA12.

On the ^{13}C NMR spectrum the peak presence at 39.56 ppm indicates that amide block is built into copolymer macromolecule.

^{13}C NMR and FT-IR analysis confirmed the assumed chemical structure of terpoly(ester-ether-amides) and terpoly(ester-aliphatic-amides). However, one should bear in mind that the terpoly(ester-ether-amide)s and terpoly(ester-aliphatic-amide)s are random block polymers.

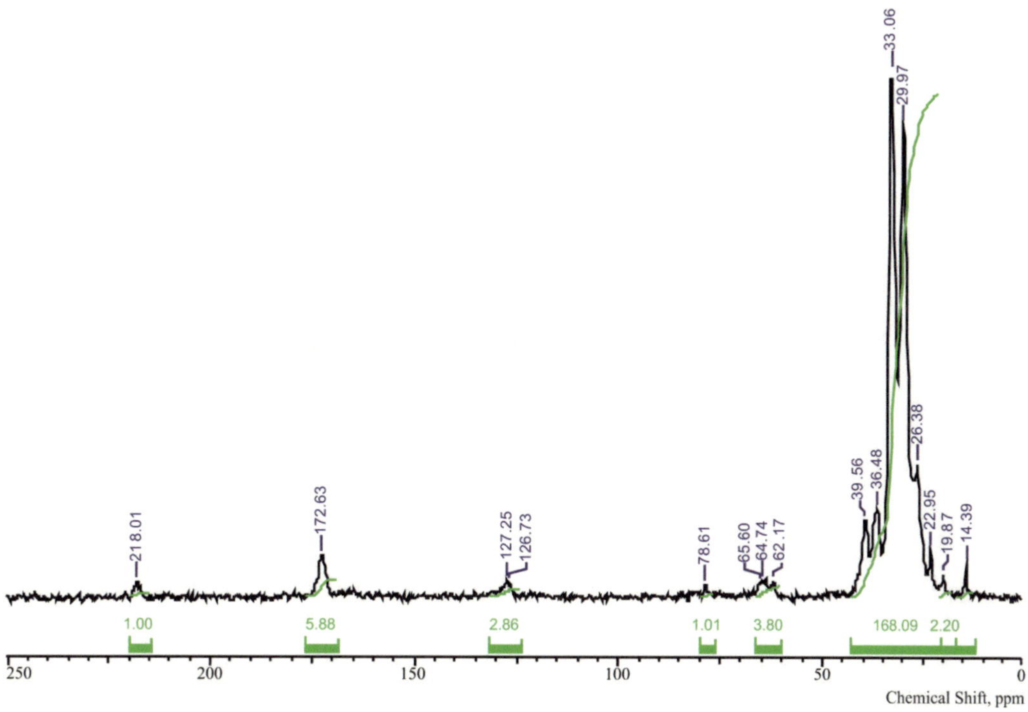

Figure 6. ^{13}C NMR spectra of terpolymer 3GT-DLAol-PA12.

Band frequency, ppm	Chemical structure
218.5	C atoms of the amide group
172-173	C atoms of the carbonyl group
126-129	C atoms in the aromatic ring
70.6	C atoms in the ether group
61-65	CH_2 groups linked to the carbonyl group through an oxygen atom
39.6	CH_2 groups bonded to a carbonyl group
36.5	CH_2 groups bonded to the C atom of the amide group
33	All other CH_2 groups in the oligoamide chain
29.8	CH_2 groups of β fragments bonded to amide or carbonyl groups or CH_2 groups of β fragments of glycol
26.9	Polyether CH_2 groups bonded to the other CH_2 groups
19 - 26	C atoms in the aliphatic chain
14.39	C18 atoms in the aliphatic chain

Table 5. Chemical shifts (ppm) of terpoly(ester-ether-amides) and terpoly(ester-aliphatic-amides).

3.4. Mechanical properties

The ability of instant recovery after the deformation is expressed by the elastic elongation(ε_s) and the area A, which is proportional to the dissipated elastic energy. It is a measure of the quality of the spatial network of the elastomer. High elastic elongation (ε_{hs}) and the area B, which is proportional to the dissipated energy, characterize the ability of a material to have delayed recovery after deformation. It is a measure of the quality of the continuous phase capable of viscoelastic response. Permanent set (ε_{ps}) corresponds to the irreversible changes that have occurred in the material under the stress, most likely related to the change in the spatial distribution of domains. The area C is proportional to the accumulated energy. Received two types of mechanical hysteresis loops. The first type was obtained by stretching a terpolymer sample from 10% to 100% at elongation growing by 10% (Figure 7). The second one was obtained by stretching at constant elongation of 100% (Figure 8).

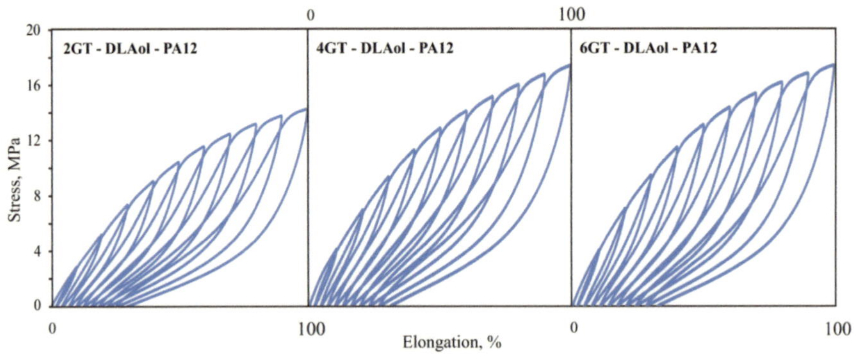

Figure 7. Mechanical hysteresis loops at elongation growing by 10% of the terpoly(ester-aliphatic-amide).

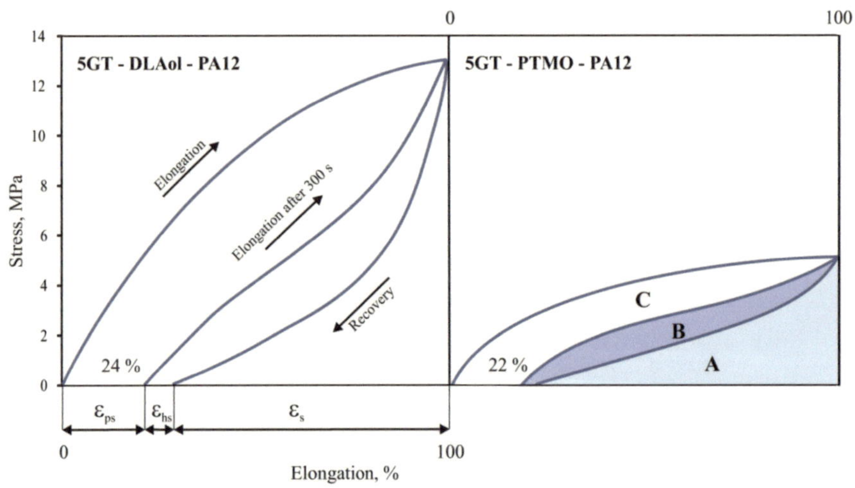

Figure 8. Mechanical hysteresis loops at a constant elongation of 100% of terpoly(ester-aliphatic-amide) and terpoly(ester-ether-amide), where the ester block was 5GT: εs – elastic elongation, εhs – delayed high-elastic elongation, εps – permanent set, A – area proportional to the dissipated elastic energy, B – area proportional to the dissipated high-elastic energy, C – area proportional to the accumulated energy.

http://dx.doi.org/10.5772/61215

In both series, the best mechanical properties have the terpolymer with the number x=4, whereas the best elastic residues exhibit terpolymers with x=5. Better mechanical properties, but worst elastic residues have terpolymers where the amorphous phase is DLAol (series II). This is probably due to the lower weight content of flexible blocks in the terpolymers. Terpolymers of this series, similar to PEE and PEA, exhibit a large part of energy accumulated during the first cycle of elongation. The probable cause is the strong interactions at the domain-matrix contact in these materials, due to hydrogen bonds, Van der Waals forces, or through Chain foldings. The mechanical hysteresis loops of terpoly(ester-ether-amide)s (series I) demonstrate that there is in these materials a small energy accumulation in the first cycle. Therefore, they are elastomers with a better mechanical shape memory than terpoly(ester-aliphatic-amide).

3.5. WAXS analysis

The crystal structure of PTT homopolymer is observed at scattering angles 2Θ of 15.08°, 16.51°, 19.17°, 23.19°, 24.39°, and 26.91°. The most intense diffraction peaks appear in doublets. The diffraction pattern of PA12 homopolymer has one wide diffraction maximum with two extreme points: 20.44° and 21.26°, which are a result of overlapping of the reflections originating from two polymorphic structures γ and α PA12. The terpolymer diffraction patterns exhibit only two reflections with the glancing angles 2Θ corresponding to the angle values on the PA12 diffraction pattern.

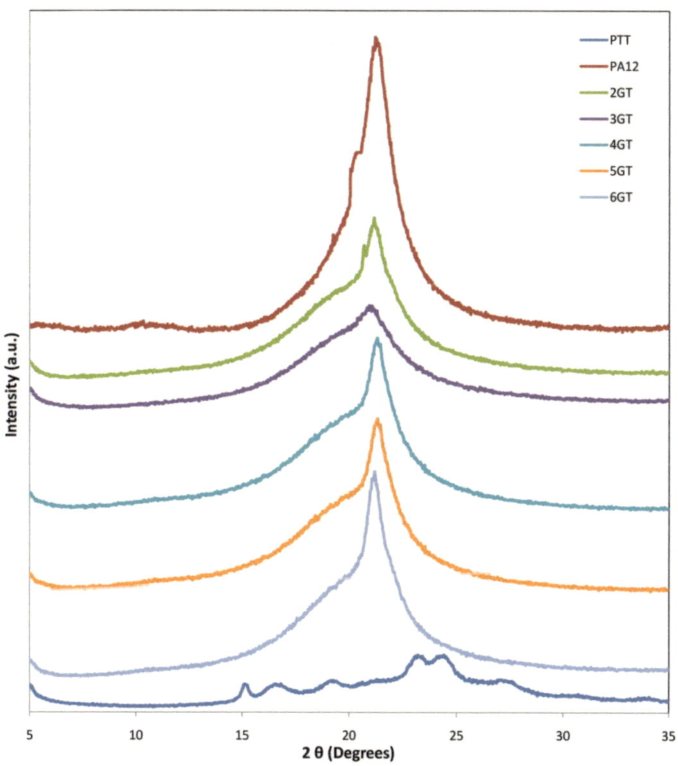

Figure 9. WAXS diffractograms of PTT, PA12, and multiblock terpolymers where the soft phase is DLA.

The qualitative analysis of the diffactograms (Figures 9 and 10) suggests that in the obtained terpolymers composed of the ester and amide hard blocks, only the amide block is responsible for the formation of the crystalline phase. For all the series, terpolymers where the ester block is trimethylene terephtalate exhibits poor-shaped and very wide diffraction maximum. This may indicate the weakest capacity for crystallization of these polymers. There are fewer polymorphic structures γ of PA12 in terpolymers where the matrix is DLAol than in those where the matrix is PTMO.

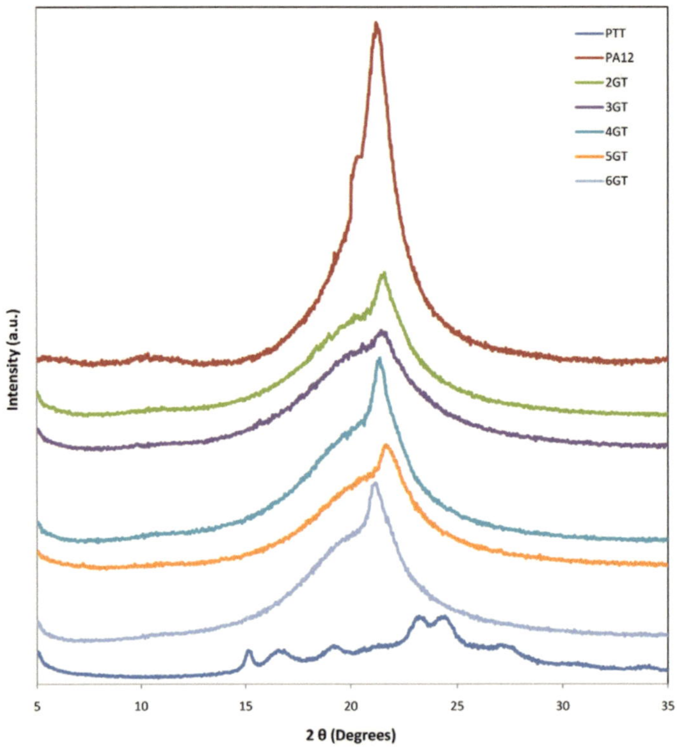

Figure 10. WAXS diffractograms of PTT, PA12, and multiblock terpolymers where the soft phase is PTMO..

3.6. Thermal properties

The terpolymer samples was heated, cooled, and reheated in the temperature range from -90°C to 250°C. The DSC curves of multiblock elastomers can be divided into two parts. The trend of the first part, which is in the low-temperature range of -90°C < T < 25°C, characterizes the processes caused by the changes in the soft phase. The trend of the second part of these curves, above 25°C, characterizes the thermal properties of the hard phase. The glass transition temperature T_g, change of specific heat ΔC_p, crystallization temperature T_c, enthalpy of crystallization ΔH_c, melting point T_m, and melting enthalpy ΔH_m for the soft and hard phases in both series of terpolymers was determined. The influence of chemical compositions of ester block on the thermal properties and on the values of phase change temperatures of the products are presented in Table 6 and are shown in Figures 11–16.

Sample	T_{g1}, °C	ΔC_p, J/gK	T_{m1},°C	ΔH_{m1}, J/g	T_{m2}, °C	ΔH_{m2},J/g	T_c, °C	ΔH_c, J/g	T_{m3}, °C	ΔH_{m3}, J/g
1	-72	0,18	-5	8	49	1,9	55	14,15	122	12,3
2	-73	0,28	-2	0,21	53	2,14	33	14,66	108	4,32
3	-72	0,16	-1	0,32	61	3,24	59	15,74	133	14,07
4	-73	0,29	4	5,15	53	1,37	60	13,23	127	11,63
5	-79	0,26	4	7,32	59	4,49	76	25,25	133	18,59
6	-21	0,12	-	-	57	3,637	44	14,56	122	13,93
7	-20	0,15	-	-	56	4,156	26	11,75	112	7,236
8	-21	0,13	-	-	55	2,842	53	18,4	130	17,4
9	-21	0,12	-	-	56	4,234	35	13,05	115	10,29
10	-23	0,16	-	-	56	3,349	53	17,57	129	15,67

T_{g1}, T_{m1} – glass transition and melting point temperatures, respectively in low-temperature region; ΔC_p – heat capacity change in T_{g1}; ΔH_{m1} – heat of melting at T_{m1}; T_{m2}, T_{m3}, T_c – melting point temperatures and crystallization temperatures, respectively in high-temperature region, ΔH_c - crystallization heat in T_c; ΔH_{m2} – heat of melting at T_{m2}; ΔH_{m3} – heat of melting at T_{m3}

Table 6. The DSC study results for terpolymers.

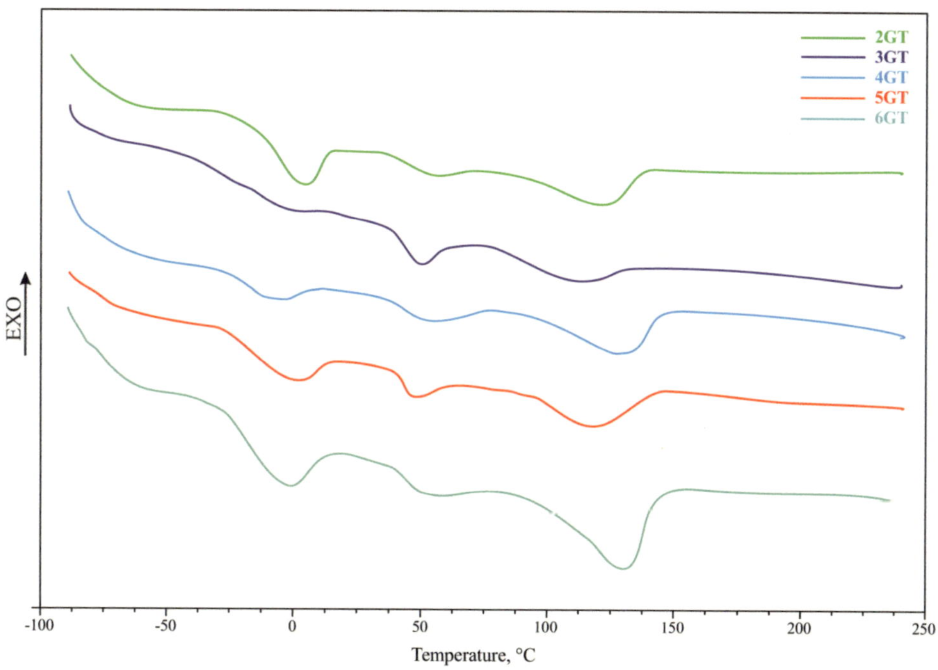

Figure 11. The first heating scans of the terpolymers of series I.

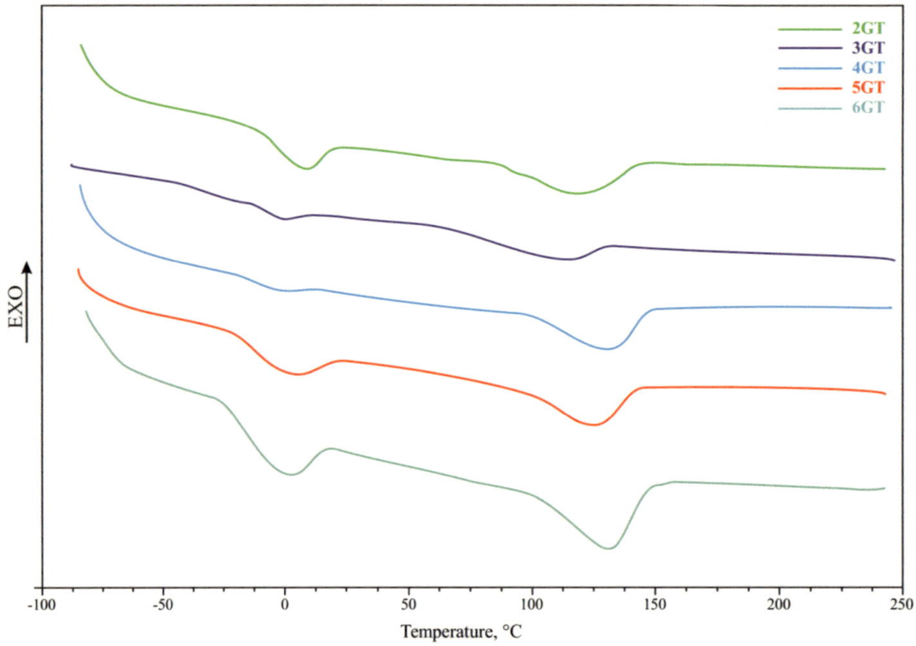

Figure 12. The second heating scans of the terpolymers of series I.

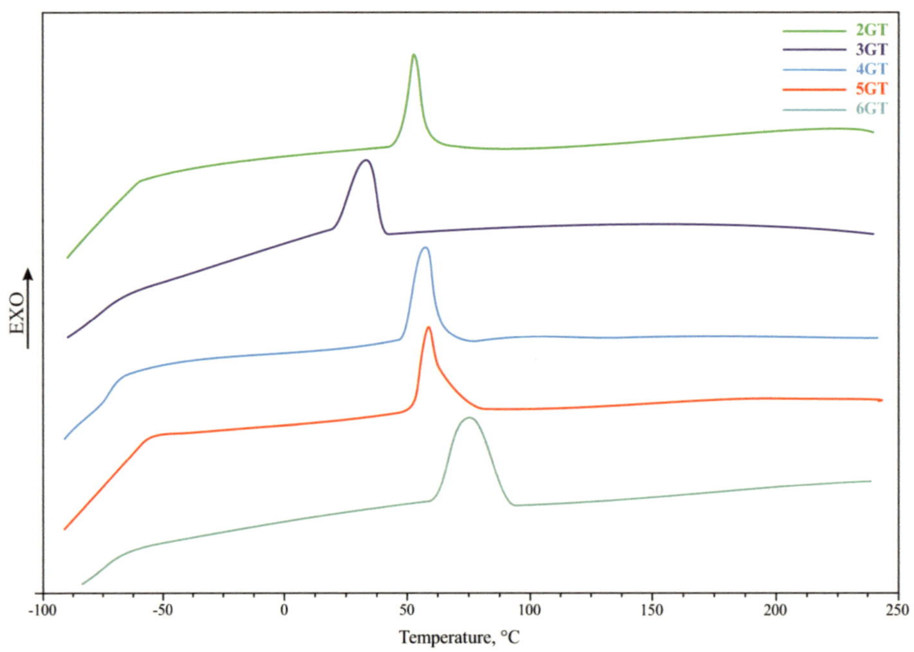

Figure 13. The cooling scans of the terpolymers of series I.

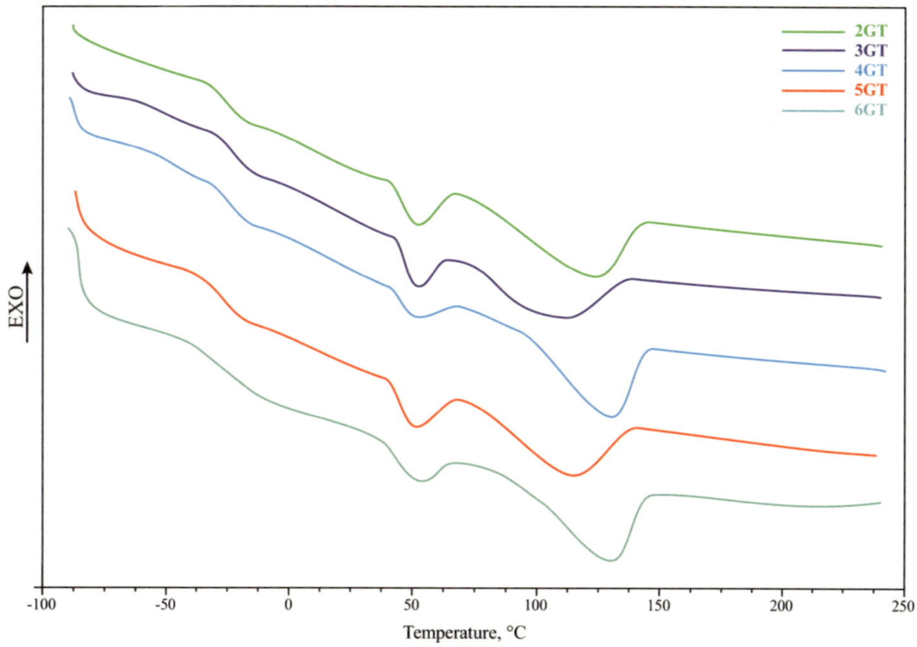

Figure 14. The first heating scans of the terpolymers of series II

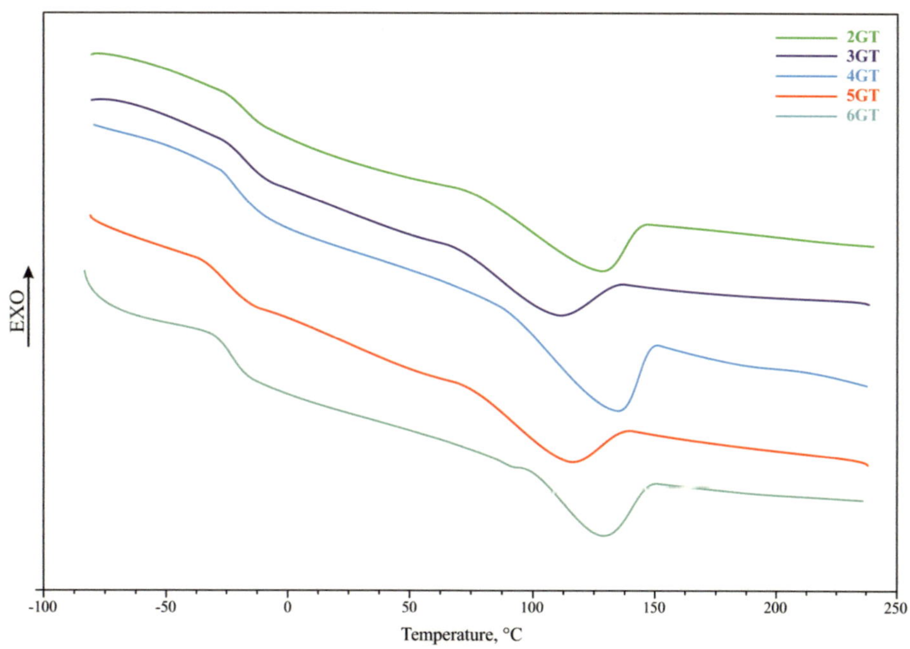

Figure 15. The second heating scans of the terpolymers of series II.

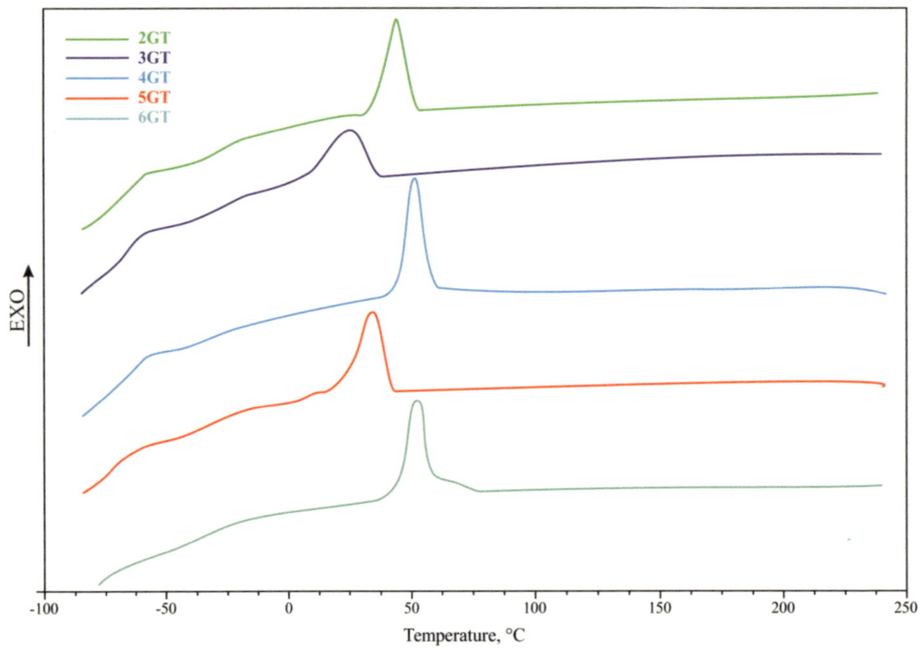

Figure 16. The cooling scans of the terpolymers of series II.

Characteristic for xGT-PTMO-PA12 terpolymers is the constant value of the glass transition temperature ($T_{gPTMO} \approx$ -72°C) in the low-temperature region, which is independent of the ester block chemical structure. The glass transition of PTMO homopolymer is -90°C and differs from the obtained terpolymers by 20°C. For the xGT-DLAol-PA12 terpolymers, difference between Tg of flexible block ($T_{gDLAol} \approx$ 61°C) and terpolymers is even greater and amounts 40°C. The immobilization of the chain-ends of the flexible block by the chemical bond enhances its T_g of about 5°C–10°C, and the interaction of the dispersed phase on this block by a further 5°C. The difference between the homopolymer glass transition temperatures and obtained terpolymers received up to 20°C and 40°C, therefore, cannot be explained by immobilization of the chain-ends or by interphase interactions. Probably, further increase in T_{g1} is responsible for the terpolymers dissolution of the short ester sequence in the soft phase. The poor-shaped melting endotherm is observed in the low-temperature region in terpolymers of series I. It determined the heat of fusion of the soft phase. With the increase of the number of carbons separating the terephthalate groups in the ester block of terpolymers also increased the melting point temperature of the crystalline fraction of PTMO blocks. This shows that the purity of the PTMO soft phase increases. In the high-temperature part of the DSC curves two melting endotherms are observed. The first thermal effect, which is called annealing endotherm, is characteristic for many polymers crystallized from the melt and it disappeared during the second heating. It is understood that it accounts for the melting of defected, small crystallites and is associated with the heat of dispersion of the mesomorfic aggregates occurring in terpolymers. The melting point temperature increased with the increasing amounts of the carbons separating the terephthalate groups in the ester block of terpolymers. For the samples with odd number of carbons x in the ester block, T_m is by 10°C–15 °C lower than in the case of the other terpolymers (principle of parity). The same regularity is observed during cooling of the materials. The difference in the chemical structure of the esters block practically does not influence on the maximal temperature range of application, which is about 200°C.

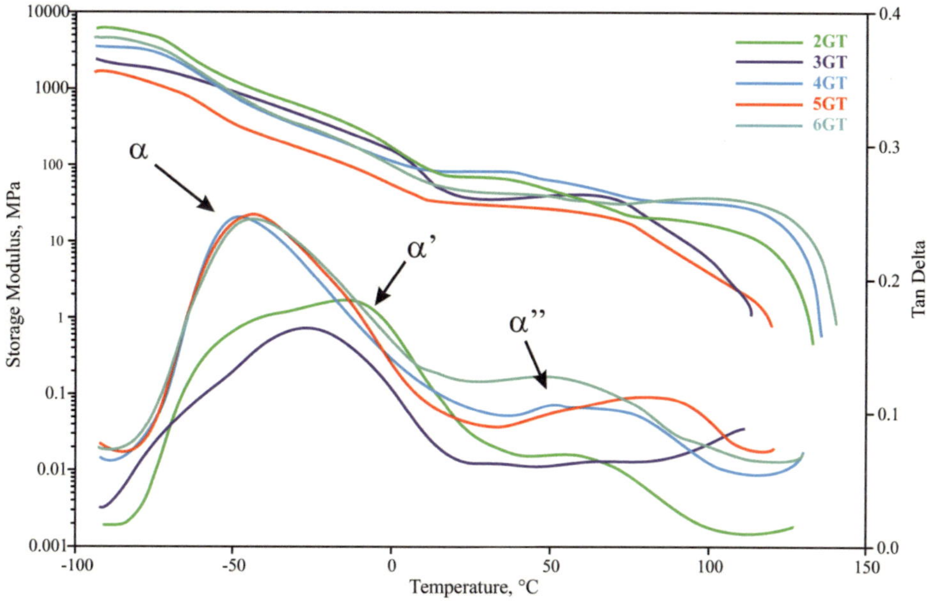

Figure 17. DMTA analysis of the terpolymers of series I.

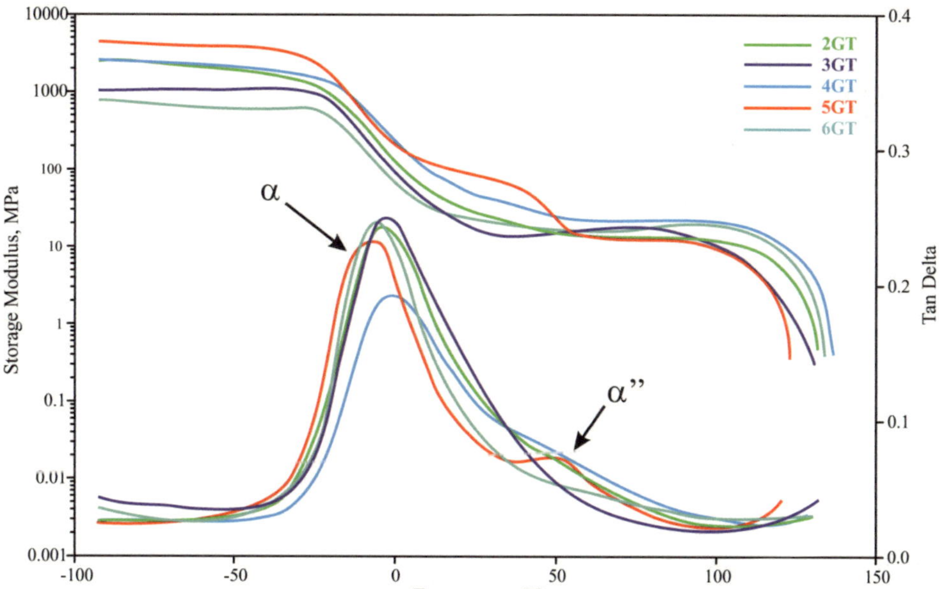

Figure 18. DMTA analysis of the terpolymers of series II.

The effect of temperature on the dynamic mechanical properties of TPEEA depending on the chemical structure of ester block was presented in Figures 17 and 18. The obtained temperature spectra are the curves being characteristic for the thermoplastic elastomers. The spectra of the storage modulus have three temperature regions, in which the courses E'=t(T) differ significantly. In the temperature region from 100°C to -70°C (PTMO) or -20°C (DLAol), the obtained TPEEAs exhibited a constant, characteristic for the glass state, the value of storage modulus above 1 GPa. In the region from -70°C (PTMO) or -20°C (DLAol) to 10°C, a decrease of modulus was observed that was caused by the appearance of viscoelasticity relaxation processes. A further increase of temperature caused the occurrence of a wide "plateau" of elastic state, which at a temperature of 120°C terminates by a rapid lowering at the point of crystallite melting of the hard block phase. For all xGT-PTMO-PA12 samples, the tgδ curves possess a broad relaxation peak composed of two relaxation transitions α and α'. They may be linked with the glass transition temperatures for region composed of pure PTMO and mixture of PTMO/xGT of the amorphous phase and interphase. Maximum α'' is a result of the relaxation effect of the amorphous phase of PA12. xGT-DLAol-PA12 terpolymers has one narrow and high damping peak. Probably, in these terpolymers the interphase is smaller because of weak mixing of ester blocks with DLAol matrix.

4. Conclusion

The synthesis, structure, and properties of poly[(multi-methylene terephthalate)-block-(oxytetramethylene)-block-(laurolactam)] and poly[(multi-methylene terephthalate)-block-(linoleic alcohol dimer)-block-(laurolactam)] terpolymers have been reviewed in this chapter. The influence of the number of carbons separating the terephtalate groups of the ester groups in the benzene ring of other blocks and on the properties and structure of these elastomers have been evaluated.

A series of new thermoplastic block elastomers was prepared by melt polycondensation. Synthesis of poly(ester-b-ether-b-amide) terpolymers was a two-step process in the presence of a titanate catalyst. The first step was the transesterification reaction of dimethyl terephthalate and glycol and the esterification reaction of α,ω-dicarboxylic oligo(laurolactam) with oligo(oxytetramethylene)diol or linoleic alcohol dimer, which is taking place simultaneously in another reactor. The second step was polycondensation reaction of two previously prepared intermediate compounds. A detailed description of this synthesis is given in previous papers [36, 37].

^{13}C NMR and FT-IR methods were used to confirm terpolymers assumed chemical structure. Obtained copolymers had all characteristic FT-IR bands for esters, aliphates or ether, and amides. There is no in obtained terpolymers a strong, wide band for the O-H stretch in the region 3300-3000 cm^{-1}, which is observed in FT-IR spectra of pure dicarboxylic oligoamides. It proves that these blocks are built into the copolymer macromolecule. This conclusion is confirmed by the peak presence at 39.56 ppm in ^{13}C NMR spectra.

The degree of phase separation of soft phase and the degree of crystallinity of hard phase was determined by DSC analysis. The interphase size was also estimated, and the occurrence of the semicrystalline structures was noticed. The DMTA and WAXS methods supplemented these data. It was found that synthesized copolymers exhibit a multiphase (crystalline-amorphous) physical structure. The amorphous phases (matrix) are composed of the flexible blocks PTMO or DLAol that are contaminated by short ester sequences. The crystalline phase (domains) is composed of the hard blocks PA12 and is disordered by admixtures of the xGT blocks. The number of carbons separating the terephthalate groups in the ester block has little effect on the physical properties of the terpolymers obtained by slightly increasing the amorphous phase.

It has been concluded that in both series the best elastic residues have the terpolymer with the number of carbons x=5. Probably the interphase of these samples is large and well shaped. Better elastic properties are exhibited by terpolymers of series I, where the soft phase is PTMO because these block is more flexible (large capability for motion and rotation of ether bond).

Obtained terpolymers exhibit unique properties, such as low glass transition temperature, a wide temperature range of application, fast cristallization, good mechanical properties, including good elasticity, thermal stability, and thermal and chemical resistance, and may find application in practice.

Author details

Joanna Rokicka* and Ryszard Ukielski

*Address all correspondence to: joanna.rokicka@zut.edu.pl

West Pomeranian University of Technology Szczecin, Poland

References

[1] Salamone J.C., editor. Polymeric Materials Encyclopedia. New York: CRC Press; 1992.

[2] Holden G., Legge N.R., Quirk R., Schoeder H.E., editors. Thermoplastic Elastomers. Munich: Hanser Publishers; 1996.

[3] Balta Calleja F.J., Rosłaniec Z., editors. Block copolymers. New York: Marcel Dekker; 2000.

[4] Walker B.M., Rader Ch.P., editors. Handbook of Thermoplastic Elastomers. New York: Van Nostrand Reinhold; 1988.

[5] Fakirov S., editor. Handbook of Condensation Thermoplastic Elastomers. Weinheim: WILEY-VCH; 2005.

[6] Holden G. Understanding thermoplastic elastomers. Munich: Hanser Publishers; 2000.

[7] Bhowmick A.K., editor. Current topics in Elastomer research. Boca Raton: CRC Press; 2008.

[8] Dick J.S., editor. Rubber Technology. Compounding and Testing for Performance. Munich: Hanser Publishers; 2001.

[9] Lal J., Mark J.E., editors. Advances in elastomer and rubber elasticity, proceedings symposium. New York: Plenum Press; 1985.

[10] Harrell L.L. Macromolecules. 1969;2(6).

[11] Ng H.N., Allegrazza A.E., Seymour R.W., Cooper S.L. Polymer. 1973;15.

[12] Miller J.A., Shaow B.L., Hwang K.K.S., Wu K.S., Gibson P.E., Cooper S.L. Macromolecule. 1985;18(1).

[13] White J.R., De S.K., editors. Rubber Technologist's Handbook. Shawbury: Rappra Technol. Ltd; 2001.

[14] Rzymski W.M., Radusch H.J. Nowe elastomery termoplastyczne. Polimery. 2005;4.

[15] Mark J.E., Erman B., Eirich F.R., editors. Science and technology of rubber. San Diego: Academic Press; 1978.

[16] Bayer O. Mod. Plast. 1947;24.

[17] Mark H., Flory P.J., editors. High Polymers. New York: John Wiley & Sons; 1964.

[18] Van der Schuur M.J., Gaymans R.J. Polymer. 2007;48.

[19] Yang I.K., Tsai P.H.. Polymer. 2006;47.

[20] Huang J.J., Keeskkula H., Paul D.R. Polymer. 2006;47.

[21] Krijgsman J., Husken D., Gaymans R.J. Polymer. 2003;44.

[22] Armstrong S., Freeman B., Hiltner A., Baer E. Polymer. 2012;53.

[23] Nery L., Lefebvre H., Fradet A. Journal of Polymer Science: Part A. 2005;43.

[24] Mateva R., Zhilkova K.R., Zamfirova G., Diaz-Calleja R., Garcia-Bernabe A. Journal of Polymer Science: Part B. 2010;48.

[25] Sugi R., Hitaka Y., Sekino A., Yokoyama A., Yokozawa T. Journal of Polymer Science: Part A. 2003;41.

[26] Ukielski R. Polimery. 1997;42.

[27] Ukielski R., Rokicka J. Przemysł Chemiczny. 2010;12.

[28] Ukielski R. Elastomery multiblokowe terpoli(estro-b-etero-amidy): Synteza, struktura, właściwości [thesis]. Szczecin: West Pomeranian University of Technology Press; 2000.

[29] Ukielski R. Polimery. 2002;47.

[30] Ukielski R. Macromolecular Materials and Engineering. 2001;286.

[31] Armelin E., Franco L., Rodriguez-Galan A., Puiggali J. Macromolecular Chemistry and Physics. 2002;203.

[32] Wu C.S. eXPRESS Polymer Letters. 2012;6(6).

[33] Wilhelm M., Neidhofer M., Spiegel S., Spiess H.W. Macromolecular Chemistry and Physics. 1999;200.

[34] Goodman I., Maitland D.J., Kehayoglou H. European Polymer Journal. 2000;36.

[35] Vlahov G. Progress in Nuclear Magnetic Resonance Spectroscopy. 1999;35.

[36] Ukielski R. Polimery. 1995;40.

[37] Ukielski R. Polimery. 1996;41.

Chapter 3

Plasticization and Morphology Development in Dynamically Vulcanized Thermoplastic Elastomers

Shant Shahbikian and Pierre J. Carreau

Additional information is available at the end of the chapter

http://dx.doi.org/10.5772/61414

Abstract

Dynamically vulcanized thermoplastic elastomers constitute one of the main categories among various types of thermoplastic elastomers (TPEs). Due to the commercial importance of this particular group of TPEs, tremendous efforts have been dedicated to improve the understanding and control the phase morphology development. The ultimate goal is to obtain materials with improved physical and mechanical properties. As in other polymeric compounds, the parameters during the mixing stage have a significant influence on the final morphology of dynamically vulcanized blends. Furthermore, the phase morphology and, therefore, the distribution of elastomeric domains in the thermoplastic phase are also strongly dependent on the formulation. This chapter discusses the main important processing factors and, more specifically, highlights the effects of plasticization and curing on the morphology development of dynamically vulcanized thermoplastic elastomer blends. The following text provides fundamental information on how one should take into consideration each parameter affecting the morphology of nonreactive and reactive elastomer/thermoplastic blends.

Keywords: Thermoplastic elastomer, dynamic vulcanization, plasticization, morphology, rheological properties

1. Introduction

Thermoplastic elastomers (TPEs) represent a large group of polymeric materials that are melt processable, similar to regular thermoplastics, and they exhibit rubber-like behavior identical to that of cross-linked elastomers. This special characteristic of TPEs is mainly due to the presence of thermoreversible cross-links, which are broken during the melt processing step under high shear and elevated temperature, and formed once again when melt processing is over and the compound reaches ambient temperature. The main concept behind the thermoplastic elastomers and, therefore, thermoreversible cross-links is the simultaneous presence of phase-separated hard and soft segments in the compound. Although both phases contribute to the overall physical and mechanical properties of the final product, some specific properties are usually associated with one phase or the other. Below the melting temperature of the hard phase, this phase usually provides the strength, stiffness, and chemical resistance of the material. On the other hand, the soft one acts as an elastomer providing the flexibility and the elastic nature by controlling the hardness, the compression, and the tensile sets. Furthermore, the soft phase dictates the lower service temperature limit of the product.

Commercially available TPEs can be generally classified into the following four groups presented in Table 1.

I	II	III	IV
Block copolymers	**Random copolymers**	**Ionomers**	**Blends of soft elastomer and hard thermoplastic**
Triblock or segmented block copolymers:			*Simple (nonreactive) or dynamically vulcanized blends (DV)*
SBS SEBS TPU COPE COPA and many more	EPR Ethylene-α Olefin Propylene-α Olefin and many more	Ethylene-methacrylic acid Ethylene-acrylic acid Butadiene-acrylic acid and many more	EPDM/PP NBR/PP NBR/PVC IIR/PP and many more

SBS: Poly(styrene-butadiene-styrene); SEBS: Poly(styrene-ethylene-butadiene-styrene); TPU: Thermoplastic polyurethene; COPE: Copolyether-ester elastomers; COPA: Copolyamide elastomers; EPR: Ethylene propylene rubber; EPDM: Ethylene propylene terpolymer rubber; PP: Polypropylene; NBR: Nitrile butadiene rubber; PVC: Polyvinyl chloride; IIR: Butyl rubber (isobutylene-isoprene rubber).

Table 1. Classification of thermoplastic elastomers (TPEs) along with few examples.

According to the general categories of TPEs shown in Table 1, the aforementioned properties associated with TPEs could be obtained through numerous paths. In the first three groups, the material mainly consists of blocks or grafted segments of soft and hard constituents through polymerization [1,2]. In addition to these large varieties of TPEs obtained through polymerization, mechanical blending of conventional elastomers with plastics provides another accessible route toward the production of TPEs. In this approach, the polymeric constituents are blended in a conventional melt mixing equipment, such as a twin-screw extruder. The result is often an immiscible blend, particularly due to the fact that the majority of polymeric pairs of large molecular weight and low interactive forces tend to form phase-separated microstructure. Interestingly, the phase-separated nature of these blends is what it is needed to exhibit properties associated with TPEs.

Although some simple nonreactive elastomer/thermoplastic blends, such as ethylene-propylene rubber/polypropylene (EPR/PP), have gained tremendous attention for their use in the automotive industry [3], the nonreactive blends have generally poor elastic recovery and poor hydrocarbon fluid resistance in comparison to their reactive counterparts [4]. Furthermore, the morphology of an immiscible simple blend is prone to change during the reprocessing and downstream operations. These issues may extremely affect the physical and mechanical properties of these types of TPEs. Consequently, simple blending of polymeric constituents is usually not sufficient to guarantee a permanent morphological feature and, therefore, stable mechanical properties regardless of the processing history of the blend. To overcome this issue, chemically cross-linking of polymeric phases is required in order to stabilize the morphology. In this case, the reactive blends of elastomer/thermoplastic blends are generally obtained through dynamic vulcanization (DV) where the elastomer phase is selectively cross-linked in the presence of a curing system. These blends are commonly known as thermoplastic vulcanizates (TPVs). The origin of DV process comes from the work of Gessler and Haslett on polypropylene and chlorobutyl rubber, where the elastomer was cross-linked in the presence of zinc oxide [5]. Fischer [6-9] further pushed the boundaries of our understanding by working on dynamically and statically cross-linked elastomer blends with polyolefins. The work of Fisher on dynamic vulcanization was further extended by complementary investigations done by Monsanto [10-13].

In both simple and dynamically vulcanized blends, the thermoplastic elastomer behavior is usually obtained for high concentration of the elastomer phase, e.g., higher than 50 wt%. This single argument on the concentration of the elastomeric component has a huge consequence on the morphology. The simple blends with high elastomer content tend to form a co-continuous morphology, where both elastomer and thermoplastic constituents are interconnected throughout the whole bulk of the material (Fig. 1a). However, in dynamically vulcanized blends, the elastomer (usually the major and initially part of the co-continuous phase) becomes discontinuous and dispersed during dynamic vulcanization (Fig. 1b). Eventually, the thermoplastic becomes the continuous phase surrounding the cross-linked elastomer particles. This morphology transformation is known as *phase inversion*.

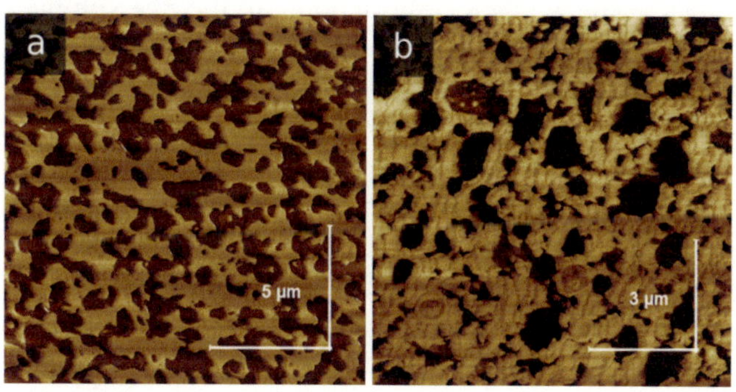

Figure 1. AFM phase micrographs of: (a) nonreactive simple, and (b) dynamically vulcanized EPDM/PP 50/50 (wt/wt %) blends (from [14])

By far, the importance of the phase-separated structure and the role of morphology have been clearly emphasized. For one familiar with polymer blending technology, the fine control of phase morphology even in a simple nonreactive blend composed of merely two components represents a huge challenge. Moreover, the presence of a complex flow field in industrial mixing equipment with simultaneous breakup and coalescence of the dispersed phase creates a far more complicated environment for comprehending the phase morphology development. The level of complexity may even increase when additional components such as processing aids, fillers, plasticizers, and curing system are added. This is commonly the case in the TPV industry where the fine-tuning of the final properties is achieved by combining several different reactive and nonreactive additives.

2. Plasticization

The use of plasticizers in both rubber and thermoplastic elastomer industries is a well-established technology [15]. Despite the overall processing cost reduction of the final product in the presence of a plasticizer, some technical aspects of a plasticized compound can also be improved. For instance, plasticizers have been used to improve the low temperature mechanical properties to reduce the hardness and acts as a dispersion aid for fillers and additives [16]. It further improves the resistance to oil swell, heat stability, hysteresis, permanent set, elastic recovery, as well as the melt processability and the final appearance of the compound [13].

Several theories such as gel [17,18], lubricity [19-21], and free volume [22] theories have so far been developed and further extended to explain different mechanisms involved during plasticization. In a rather general way, the lubricity theory considers that a plasticizer reduces the intermolecular friction between polymer chains, which is originally considered as its source of rigidity. It acts as lubricant and reduces the resistance to sliding between molecules. In the gel theory, the polymer molecules are considered to form a tridimensional structure held by loose attachments along their chains. According to this theory, the stiffness of the polymer is mainly due to the presence of this tridimensional structure. In such systems, plasticizers act in favor of reducing the number of attachments between polymer molecules and, therefore, enabling the molecules to change their conformation. In the free volume theory, the friction between polymer chains is attributed to the volume between molecules. The free volume in polymers is essentially considered as the required space for chain mobility. By increasing the temperature in a nonplasticized system, the chain mobility increases as a consequence of an increase in free volume. On the other hand, the shrinkage of the free volume with decreasing temperature may reach to critical level where only limited free space is available for polymer chains to have large segmental motion. This critical temperature is known as the glass transition temperature (T_g). Therefore, polymers below Tg behave as solid glassy materials, whereas at temperatures above Tg they possess rubber-like properties. According to this theory, the free volume in a polymer may be increased through different paths [23]: (1) by increasing the temperature (as mentioned earlier), (2) by lowering the molecular weight of the polymer resulting in an increased concentration of end groups, (3) by incorporation and/or increasing the length of side chains in the polymer, (4) by incorporating segments with low steric hindrance and low intermolecular interaction along the polymer chains, (5) by adding lower molecular weight compounds with lower *Tg*, which are compatible with the polymer. Consequently, a decrease in the glass transition temperature and an increase in the mobility of polymeric chains as a function of the plasticizer concentration could be readily associated with increases in the free volume.

Once a plasticizer is incorporated and completely dissolved at the molecular level in a desired polymer, the combination of all the aforementioned theories provides an extensive insight into the plasticization mechanisms. However, dissolution of a plasticizer into polymer and the compatibility may largely affect the efficiency of plasticization in both the short and the long term. The compatibility issue is the principal factor in determining the proper plasticizer for a given polymer. Generally, a compatible polymer/plasticizer pair is by nature a homogeneous mixture.

The incorporation of a plasticizer during an industrial compounding process is usually achieved by the expenses of mechanical energy. However, afterward the stability of the compound is directly related to the thermodynamic phase equilibrium between the components. As a result, the widely used compatibility concepts are, therefore, based on the theories of polymer solutions. In a pair of plasticizer/amorphous polymer with an upper critical solution temperature (UCST) phase diagram, the homogeneity of the mixture increases with temperature. At lower temperatures, a mixture containing around 15 to 40 vol% of plasticizer (a typical range in TPVs) may be phase-separated into polymer-rich and plasticizer-rich domains. In certain cases, a lower critical solution temperature (LCST) may also be observed. In contrast to UCST phase diagrams, those systems displaying LCST tend to phase-separate at higher temperature. The complexity of the phase diagram is normally increased when instead of an amorphous polymer; a semicrystalline polymer is to be plasticized. The crystallization process of the polymer tends to be the factor that complicates the formation of a homogeneous polymer/plasticizer mixture. At temperatures high enough to obtain a homogeneous mixture between the amorphous fraction of the polymer and the plasticizer, but low enough that crystallinity still prevails, the presence of local crystal network affects negatively the compatibility and in certain conditions it results in migration of the plasticizer. This reasoning on the effect of crystallinity brings us to a general discussion regarding the flexibility of polymer chains. Flexible chains are known to dissolve more easily in a plasticizer. This mainly is due to lower energy requirement in separating polymer chains, and an easier diffusion of chains in the plasticizer. This process usually increases the mixing entropy and facilitates the mixing. This again demonstrates that the dissolution of crystalline or cross-linked polymer chains in a plasticizer is a laborious task and, therefore, the compatibility between these pairs is much less than the compatibility between flexible amorphous polymer chains and a plasticizer. Beside chain flexibility, crystallinity, and cross-linking, some other criteria have to be as well considered. Some are based on the polarity differences between plasticizer/polymer pairs. Highly polar polymers do not usually dissolve in nonpolar solvents and the other way around. However, some exceptions to this rule have already been observed.

Up to this point, most of the discussion surrounding the compatibility issue was qualitative. However, to choose a proper plasticizer one should estimate the compatibility based on a measurable value derived from the Flory-Huggins theory [24,25]. The miscibility criterion in this theory is determined by the Gibbs free energy of mixing, which is a combination of the enthalpy and entropy of mixing:

$$\frac{\Delta G^M}{RTV} = \left[\frac{\phi_1}{V_1} ln\phi_1 + \frac{\phi_2}{V_2} ln\phi_2\right] + \left[\frac{\phi_1\phi_2}{V_1}\chi_1\right] \tag{1}$$

T	Temperature, [K]
R	Universal gas constant, [J/K.mol]
V	Total volume, [m^3]
V_1, V_2	Molar volumes of plasticizer and polymer, respectively, [m^3/mol]
ϕ_1, ϕ_2	Volume fractions of plasticizer and polymer, [-]
χ_1	Interaction parameter, [-]

Table 2. Description of parameters of eq. 1

The term in the first bracket represents the entropy change of mixing, whereas the second one represents the enthalpy change. In a polymer/plasticizer mixture, an instantaneous miscibility (or complete compatibility) is achieved when the Gibbs free energy of mixing is negative. This can only be achieved if the enthalpy contribution is small enough in comparison to the entropic one. Therefore, the interaction parameter (χ_1), which represents the enthalpic contribution, is an important feature in estimating the compatibility. It characterizes the difference between the interaction energy when a plasticizer molecule is immersed in a neat polymer, versus when it is immersed in a neat plasticizer. The upper limit for χ_1 to obtain a compatible pair is 0.5. For values greater than 0.5, incompatibility and phase-separation may be observed. Although values for χ_1 in different polymer/solvent (plasticizer) systems can be found in the literature and in several polymer handbooks [16,26,27], in practice the most useful parameter used to estimate the molecular interactions between two components is the solubility parameter (δ). This parameter is the square root of the cohesive energy density. This is the energy that has to be given to a system of pure liquid to extract a molecule from the liquid state. Therefore, it is proportional to the interaction energies between similar molecules. The change in the internal energy of mixing can be related to the solubility parameter based on the following equation [28]:

$$\Delta U^M = \Delta H^M\Big|_{Const.\ Volume} = V\phi_1\phi_2\left[\delta_1 - \delta_2\right]^2 \tag{2}$$

The combination of eqs. 1 and 2 for the enthalpy change of mixing results in an equation relating the Flory-Huggins interaction parameter to the solubility parameters of the polymer and the plasticizer:

$$\chi_1 = \frac{V_1}{RT}(\delta_1 - \delta_2)^2 \tag{3}$$

V_1 Molar volume of plasticizer, $\left[\mathrm{m}^3/\mathrm{mol}\right]$

A more appropriate relationship documented in the literature for a mixture of polymer/low molecular weight liquid (plasticizer) is the following:

$$\chi_1 = 0.34 + \frac{V_1}{RT}(\delta_1 - \delta_2)^2 \tag{4}$$

Earlier, the limit of compatibility between polymer/plasticizer pairs was determined by a critical value (χ_1=0.5). According to this criterion, compatibility can be approximately estimated by matching the solubility parameters. A typical low molecular weight liquid (plasticizer) may have a molar volume around 100 to 400 cm^3/mol. Accordingly, this translates into a critical difference between the solubility parameters around 0.48 to 0.97 (cal/cm^3)$^{0.5}$ (based on eq. 4). As an example, for natural rubber with solubility parameter of 8.30 (cal/cm^3)$^{0.5}$ [29], a proper plasticizer shall possess a solubility parameter between 8.3 ± 0.97 (cal/cm^3)$^{0.5}$. This simple method is widely used in industry and, although it is considered as a rapid screening technique to identify a proper plasticizer, there might be cases as we will demonstrate shortly that the prediction is not quite exact.

A more appropriate approach for compatibility studies in terms of solubility parameter is the use of the three-dimensional solubility components proposed by Hansen [30]. In this approach, the individual solubility parameter for each and every phase involved in the system is composed of contributions from van der Waals dispersion forces (δ_d), dipole–dipole interaction between molecules (δ_p), and the contribution from hydrogen bonding (δ_h):

$$\delta = (\delta_d^2 + \delta_p^2 + \delta_h^2)^{1/2} \tag{5}$$

The advantage of using the three-dimensional components of the solubility parameter is its ability in distinguishing between different chemical interactions, which might be present. For instance, two substances may have exactly the same overall solubility parameter, but with different proportions of $\delta_d/\delta_p/\delta_h$ components. Ethylene carbonate and methanol both possess quite similar overall solubility parameters, around 29.0 MPa$^{1/2}$ or 14.2 (cal/cm^3)$^{0.5}$ [31]. However, comparing their corresponding Hansen solubility components, one could clearly understand their difference. Ethylene carbonate with (δ_d=18.0; δ_p=21.7; δ_h=5.1 MPa$^{1/2}$) and methanol with (δ_d=14.7; δ_p=12.3; δ_h=22.3 MPa$^{1/2}$) have a huge difference in terms of hydrogen-bonding interactions [31]. As a result, if these two are to be in contact with a polymer, the difference between their solubility capabilities could be readily predicted through the difference in their three-dimensional values in the $\delta_d,\delta_p,\delta_h$ coordinates. The three-dimensional distance between two substances is usually calculated based on the following equation:

$$D = (4\times\left(\delta_{d,1}-\delta_{d,2}\right)^2+\left(\delta_{p,1}-\delta_{p,2}\right)^2+\left(\delta_{h,1}-\delta_{h,2}\right)^2)^{1/2} \tag{6}$$

A smaller distance is an indication of similarity and thermodynamic compatibility between two molecules. In the case of the previous example, the distance between ethylene carbonate and methanol molecules is around 20.7 $MPa^{1/2}$. This shows that these two molecules are not at all similar and they will not behave in a similar manner when subjected to a polymer (regardless of the type of polymer). In a plasticizer/polymer mixture, as the distance increases, the compatibility and the solubility decreases in a way that after a certain distance known as the polymer radius the compatibility is negligible. The simplest and the most practical way to calculate the compatibility between a polymer and a plasticizer is to calculate the three-dimensional distance based on the Hansen solubility components and, then, divide D by the radius of the polymer. This ratio is known as the relative energy difference (RED). A mixture with a RED value smaller than 1 is compatible or even soluble in the best-case scenario; whereas a RED value greater than 1 is a sign of incompatibility and insolubility. This approach provides a more accurate prediction for compatibility; however, care must be taken when dealing with complex mixtures with cross-linked or crystalline polymers. These complexities reduce the radius of a polymer and affect negatively the compatibility with a plasticizer.

In industry, a more complex situation is generally encountered. A plasticizer is usually incorporated into a blend of two or more polymeric constituents. Since the majority of polymer blends are known to be immiscible due to thermodynamic limitations, a question arises regarding the distribution of the plasticizer when mixed in multiphase polymer blends. The answer to this question could simply clarify whether the properties of both phases are affected equally in the presence of a plasticizer, or only one of the phases will be largely affected due to its higher affinity with the plasticizer. To quantify the characteristics of the distribution, a quantity known as distribution or partition coefficient ($K_{A/B}$) has been widely used. This coefficient is the ratio of the weight fraction of the plasticizer in the polymer A over the weight fraction of the plasticizer in the polymer B, i.e., $K_{A/B} = w_{p,A}/w_{p,B}$. A value of $K_{A/B} = 1$ means that the plasticizer is equally distributed in both polymers; whereas, values lower than 1 mean that the plasticizer has a tendency toward polymer B. According to Mishra et al. [32], in those immiscible blends where the distribution of an additional low molecular weight component is entropically driven, the lowest free energy of mixing is usually achieved in the vicinity of $K_{A/B} \sim 1$. This implies that in a system where the plasticizer has a close affinity with both polymers, a uniform distribution is thermodynamically favored. On the other hand, in a blend where only one of the components is compatible with the plasticizer, the plasticizer will obviously migrate to that specific phase. An example of such system is the blend of poly(methyl methacrylate) (PMMA) and acrylonitrile butadiene styrene (ABS) [33]. These two polymers are mutually miscible. However, in the presence of a plasticizer composed of ethylene carbonate and propylene carbonate (1:1), a phase-separated system is obtained. Since the plasticizer mixture happens to have more affinity with PMMA, a plasticized PMMA phase and an ABS-rich phase are obtained. In this special case, both ABS and PMMA are amorphous polymers. Blends with semicrystalline polymers could behave quite differently. In a conventional compounding process, the semicrystalline polymer crystallizes upon cooling from the molten state. During this stage, the plasticizer even though miscible or compatible with the semicrystalline polymer in the molten state usually migrates to the other phase. As a result, the resulting distribution of the plasticizer in the molten state could be largely different from that of the solid state. In a blend composed of polypropylene (PP) and styrene-ethylene-butadiene-styrene (SEBS), Ohlsson et al. [34] have concluded that the distribution coefficient ($K_{PP/SEBS}$) should vary between 0.33 to 0.47 for blends with 10 to 90 wt% of polypropylene (a semicrystalline polymer). This indicates a preferential distribution of the plasticizer in the SEBS. In dynamically vulcanized blends composed of EPDM/PP in the presence of a paraffinic oil, the distribution of plasticizer has been shown to depend on the concentration and the molecular weight distribution of the polypropylene, as well as on the crystallized or molten state of the material [35]. The $K_{PP/EPDM}$ was lower or close to 1 at room temperature, whereas at an elevated temperature, i.e., 190 °C where polypropylene is molten, the distribution coefficient was larger than 1. Based on a micro-mechanical modeling approach, it has also been shown that the distribution coefficient in a nonreactive PP/SEBS and dynamically vulcanized EPDM/PP blends is in favor of the elastomeric component, but varying with the composition [36]. By considering the rigid nature and difficulty of plasticizing the styrenic blocks in SEBS and crystalline portion of polypropylene, the average corrected $K_{PP/SEBS}$ has been reported to be around 0.51 for the molten state, and 0.76 for the solid state [37]. The same ratio ($K_{PP/EPDM}$) of plasticized dynamically vulcanized PP/EPDM blends was reported to be around 0.89 in the solid state. The scanning electron microscopy (SEM) images of rapidly cooled extruded strands of nonreactive EPDM/PP blends have also provided a significant insight into the distribution of plasticizer [14]. The oriented structure of the EPDM phase in the plasticized extrudate versus the nonoriented structure observed in the nonplasticized blends was associated with the predominant presence of plasticizer in the elastomer phase. The possibility of obtaining an oriented elastomeric structure in the presence of plasticizer was explained in terms of large drop in the rheological properties of the elastomer in the presence of plasticizer to a level at which the polypropylene phase could deform the EPDM phase, which was otherwise a highly viscous and elastic material. Once again, these studies illustrate that even though the chosen plasticizer could have a high affinity with both polymeric components, the preferential distribution of the plasticizer is toward the elastomeric component. In a few studies, however, a separate phase mainly composed of plasticizer was also reported [38,39].

3. Morphology development in simple nonreactive thermoplastic/elastomer blends

It was mentioned earlier that the majority of polymer blends form immiscible systems due to thermodynamic limitations. Consequently, blending usually results in a multiphase heterogeneous morphology, which along with other properties of the constituent polymers dictates the ultimate properties of the resulting compound. Nowadays, compounding is performed in conventional melt-mixing equipment, such as twin-screw extruders, Banbury, or any other internal mixers. The morphology development during the compounding stage strongly depends on the rheological and the interfacial properties of the constituent polymers, their concentrations, the processing conditions and whether or not other additives such as plasticizers, curatives or fillers are present in the system. As a result, for a given immiscible blend composed of only two polymeric components, a wide range of morphologies can be tailored to specifically fulfill the requirements related to the end-use application of the blend. A droplet/matrix type morphology, if well designed, may improve the impact properties and the toughness of materials; the lamellar type will improve the barrier properties; the fibrillar type may improve the tensile properties and the stiffness of the material; and finally the co-continuous morphology may have combinatorial effects due to the simultaneous contributions of both phases at the same time.

Generally, compounding process in conventional melt-mixing equipment begins with the materials in their solid state, e.g., pellets, powders, or bales. These are subjected to intensive mixing conditions such as elevated temperature alongside with shear, which transforms the initial solid state to a molten liquid. As a result, materials are gradually softened, deformed, and finally become molten polymeric components with corresponding viscoelastic properties. The initial transformation of solid pellets to micron-size particles in the early stage of mixing process was shown to be achieved through sheet formation [40]. This mechanism is based on the mutual contact between the solid particles of the dispersed phase with the hot metal surface of the mixing equipment, which alongside with shear results in the formation of sheets/ribbons. The final micron-size particles are finally formed through transformation of these sheets into cylinders, which themselves are broken-up through Rayleigh instabilities. Although this explains the dispersion mechanism of the minor phase, one also needs to take into account the melting behavior of the constituent polymers in the blend. At the early stage of mixing, the melting sequence has a great importance [41,42]. In the case where the minor phase has a lower melting point than the major phase, it rapidly melts and encapsulates the major phase. However, as the temperature of the bulk increases with time during the mixing operation, the major phase begins to melt and becomes the matrix encapsulating the minor phase. As a result, in those blends with component melting characteristics far apart, the melting sequence could be crucial in determining the final morphology.

Once all polymeric components are molten, the mixed medium becomes essentially a mixture of viscoelastic fluids. The morphology refinement at this stage depends on several other parameters, such as shear and elongation rates, mixing time, rheological properties of the constituent polymers, blending composition, interfacial properties, and the presence of other solid or liquid additives (e.g., fillers or plasticizers). In dynamic mixing conditions, the flow and deformation of polymeric components are closely associated to their rheological properties. Among all the rheological properties, the viscosity ratio of the polymeric components at the processing conditions is one of the well-known factors that have been directly related to the morphology of the blend [43,44]. For viscosity ratios greater than 1 ($\eta_A/\eta_B > 1$), the size of the polymeric domains is known to increase monotonically. Furthermore, Favis and Chalifoux

[43] have observed that in contrast to Newtonian dispersions, an immiscible blend with a high viscosity ratio ($\eta_{minor}/\eta_{major}$ ~ 17) is still deformable. This essentially means that in a complex flow field, where shear and elongational flow coexist, it is still possible to deform a highly viscous dispersed phase. In the other side of viscosity ratio range, a composition dependency has been observed [43]. In the low composition range, the minimum domain size in polypropylene/polycarbonate (PP/PC) blends was reported to be reached around ($\eta_{minor}/\eta_{major}$ ~ 0.15 and below this value no significant change in the domain size was observed [43]. A similar observation where the polydispersity and the average domain size increased with the viscosity ratio has been reported in EPDM/PP blends with EPDM as the minor phase [45].

The blend composition is another major factor affecting the morphology. When two immiscible polymers are compounded, the morphology in the low composition range mainly consists of droplets of the minor phase in the matrix of the major component. The size and the polydispersity of the emulsion largely depend on the compatibility and, therefore, the interfacial tension and ratio of rheological properties between the polymeric components. As the interfacial tension gets smaller and the viscosity ratio approaches unity, finer droplets of the minor phase are usually observed [46,47]. Further increase in the concentration of the minor phase results in a coarser morphology due to coalescence, where droplets are coalesced with each other and form an emulsion of larger size droplets and increased polydispersity. The increase in the concentration of the minor phase eventually transforms the morphology into a co-continuous type. Each polymeric phase in a co-continuous structure is interconnected throughout the whole bulk of the material and, therefore, both components are expected to contribute simultaneously in the overall properties of the material. Immiscible polymer blends with highly viscous and elastic components generally show a rather wide range of co-continuous composition range. At both extremities of the co-continuity interval, the polymer that constitutes the major phase tends to encapsulate the minor phase. This results in breakdown of the co-continuous structure by transforming it into dispersed-type morphology. The aforementioned co-continuity range and the factors affecting it have major importance in thermoplastic/elastomer blends. As mentioned earlier, most reactive dynamically vulcanized thermoplastic/elastomer blends are produced from their corresponding nonreactive precursor blend at a composition range that coincides with the co-continuity interval [13]. This is mainly due to the fact that in this composition range, there is sufficient amount of elastomeric and thermoplastic components present, which eventually after dynamic vulcanization the elastomer is able to provide the flexibility and rubber-like behavior, whereas the thermoplastic phase guarantees the complete encapsulation of the elastomer phase when it is transformed into dispersed particles.

Several parameters may affect the co-continuity range. Parameters such as the viscosity of the polymeric components, interfacial tension, presence of plasticizer, and mixing time are those which have been carefully studied [48-50]. The early works in understanding the concept of co-continuity were merely concentrated in investigating the effects of concentration and viscosity ratio. Recent studies, however, did not necessarily consider the viscosity ratio as an independent parameter when evaluating the onset of co-continuity [48,51]. Basically for a given viscosity ratio, the onset and therefore the width of the co-continuity range has largely been attributed to the viscosity of the major phase (η_{major}) regardless of the viscosity of the minor one (η_{minor}); after all, it is the former that is responsible for imposing stresses on the minor phase during processing. Based on some theoretical background on the stability of deformed threadlike emulsions and their packing density, it has been shown that a larger viscosity of the major phase shifted the onset of co-continuity to a lower concentration of the minor phase component [51]. This has also been experimentally observed [49] in a blend composed of EPDM/PP, where the presence of a highly viscous and elastic EPDM in comparison to a less viscous and elastic PP resulted in an asymmetric co-continuity interval, with the onset of co-continuity shifted to a lower concentration of the less viscous PP, and higher concentration of highly viscous EPDM phase as illustrated in Fig. 2.

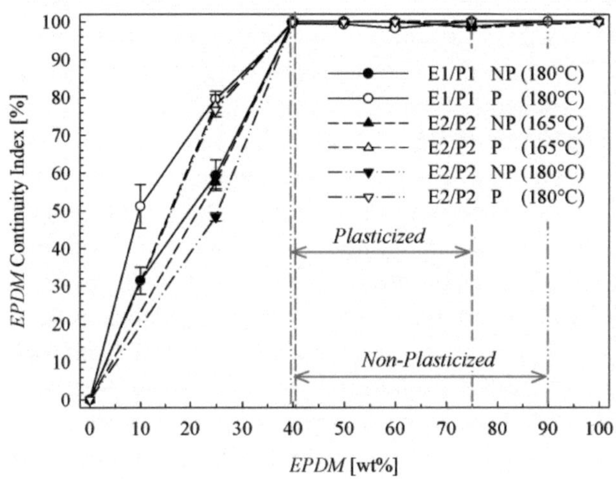

Figure 2. Continuity index of EPDM phase in both plasticized and nonplasticized blends. An asymmetric co-continuity can be clearly observed especially in the case of the nonplasticized system (from [49])

In blends with an identical viscosity ratio, the width of the co-continuity interval could be linked to the interfacial tension. Comparing two distinct immiscible blends with identical viscosity ratio, the one with a higher interfacial tension will have an onset of co-continuity shifted to higher concentrations with a narrow co-continuity interval [48]. Li et al. [52] proposed an interfacial dependent mechanism for the co-continuous morphology development. The authors proposed two distinctive coalescence mechanisms responsible for the co-continuity formation, thread–thread coalescence for low interfacial tension blends, whereas droplet–droplet coalescence mechanism for high interfacial tension blends. Accordingly, the onset and the width of the co-continuity interval is not merely associated with the rheological properties of the constituent polymers, but also to the mobility of interface and the ease of coalescence and percolation of the minor phase.

The use of low molecular weight plasticizers in an immiscible thermoplastic/elastomer blend may simultaneously affect the rheological properties of the components, the interfacial tension, and the volume of the individual phases through swelling. Based on the earlier discussion on the plasticization of thermoplastic/elastomer blends in the molten state, plasticizers have a slight tendency toward the elastomeric phase. This could clearly indicate that the rheological properties of the elastomer could be largely dropped in highly plasticized systems. Consequently, the ease of deformation and coalescence between elastomeric domains with simultaneous swelling in the presence of plasticizer could result in a coarser morphology, as shown in Fig. 3 [49]. Meanwhile in plasticized blends, a faster percolation and higher continuity index of the elastomeric phase at a lower concentration illustrated in Fig. 2 can be readily understood when the morphologies of nonplasticized and plasticized blends are compared (Fig. 3).

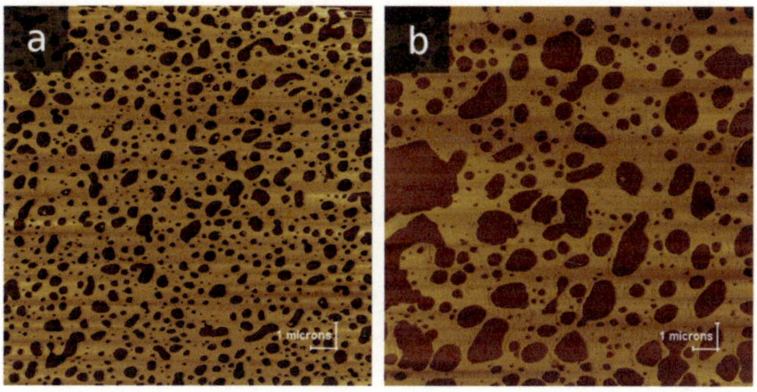

Figure 3. Coarsening effect of a plasticizer in plasticized nonreactive blends of EPDM/PP 25/75 (wt/wt%): (a) nonplasticized (b) plasticized; dark phase: EPDM; bright phase: PP (from [49])

Processing time is another factor that may or may not affect the morphology and it has been investigated by several authors [45,50,53,54]. In few studies performed on blends with high viscosity ratio ($\frac{\eta_{minor\ phase}}{\eta_{major\ phase}} \geq 1$) and low composition ranges of the dispersed phase, the effect of mixing time was shown to be insignificant regardless of the mixing equipment used, i.e., continuous twin-screw extrusion or internal mixer [53,54]. This is mainly due to the fact that the main deformation and disintegration process is considered to take place within the first few minutes of mixing. A more thorough study covering a wider composition range and blends with viscosity ratios lower or equal to unity demonstrated that the mixing time affected the morphology development, especially in the low viscosity ratio blends [50]. Bu and He [46] work investigated two blends with interfacial tensions of the same order of magnitude, but with different viscosity ratios. The morphology at longer processing times appeared not to be influenced by the viscosity ratio and only by the blend composition. However, at an early stage of mixing, the morphology seemed to depend on the viscosity ratio and whether the low or high viscous component formed the minor phase. For the low viscosity ratio system, when the low viscosity polymer formed the minor phase, the morphology consisted of droplets at low compositions and with further increases in the composition a fibrillar and eventually a co-continuous morphology was formed at a relatively low composition range. On the other hand, when the high viscosity component was the minor phase, it appeared in the form of dispersed droplets and at high concentrations it was transformed into a continuous phase. For blends with viscosity ratio in the vicinity of one, the dispersed phase in either side of the composition range appeared to be in the form of droplets up to a range where a co-continuous morphology was formed. Furthermore, regardless of the blending system and viscosity ratio, the width of the co-continuity interval was shown to decrease with mixing time. Among other processing parameters, the rotational speed (or indirectly the deformation rate) in an internal mixer has been shown not to have a significant effect on the morphology, especially in the low composition range [53,55].

In both rubber and plastic industries the compounding is usually performed in batch internal mixers or via continuous twin-screw extruders. The twin-screw extrusion process is widely known to be an excellent and versatile technique especially due to its modular design capabilities. The possibility and the advantage of designing different screw configurations generally results in intensive mixing conditions, which provides an efficient distributive and dispersive flow characteristics. Therefore, the comparison between the final morphology and its evolution in an internal mixer and a twin-screw extruder is not as straightforward as it could be imagined. However, although the types of flow fields and their intensities are largely different in those two types of equipment, the overall morphology evolution passes through the same sequences as shown by Sundararaj et al. [54] for polystyrene/polyamide (PS/PA) and polystyrene/polypropylene (PS/PP) blends. The same sequences of sheet formation, transformation into elongated domains, and eventually formation of droplets of the minor phase were observed. Regarding the final domain size of the morphological features, twin-screw extrusion has generally resulted in similar or even finer morphology in the blends [14,56,57]. In both nonplasticized and plasticized thermoplastic/elastomer blends within the co-continuous composition range, Shahbikian et al. [14] have shown that the use of twin-screw extrusion substantially refined the morphology and increased the interfacial area as illustrated in Figs. 4 and 5. The more refined structure especially in the co-continuous composition range provides a more desirable initial morphological state for further reactive and dynamic vulcanization of thermoplastic/elastomer blends.

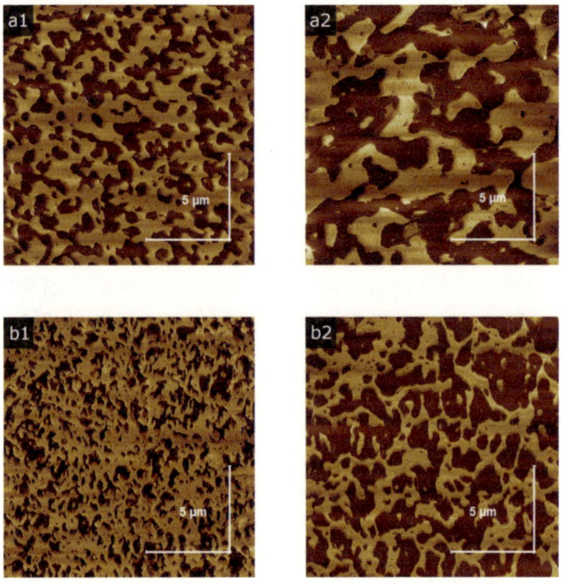

Figure 4. Atomic force microscopy images of EPDM/PP 50/50 (wt/wt%) TPOs: (a) internal mixer, (b) twin-screw extruder. (Column 1: non-plasticized, Column 2: plasticized; Dark phase: EPDM; Bright phase: PP) (from [14])

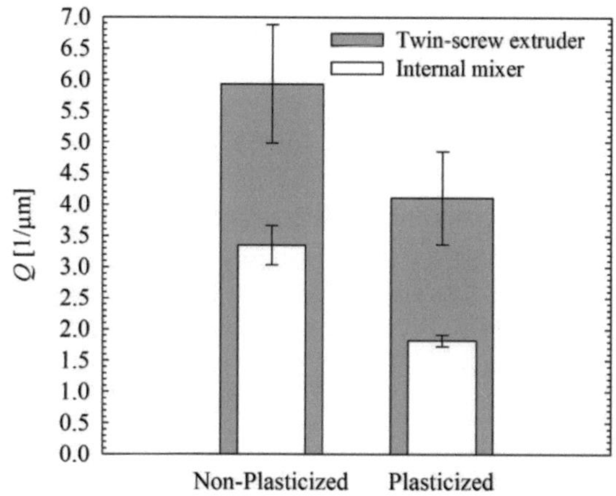

Figure 5. Specific interfacial area (Q) of EPDM/PP 50/50 (wt/wt%) prepared in internal mixer and twin-screw extruder (from [14])

4. Morphology development in dynamically vulcanized nonplasticized and plasticized thermoplastic/elastomer blends

The evolution of phase morphology in reactive blends generally takes place while the rheological, interfacial, and thermodynamic properties of the components are changing due to the chemical reaction. In the case of dynamic vulcanization of thermoplastic/elastomer blends, it is mainly the selective cross-linking reaction of the elastomeric component which influences all the aforementioned properties. The gradual formation of cross-linked elastomer network increases both the viscosity and the elasticity of the elastomer, and affects drastically the morphology development. Throughout this process, the elastomeric major phase, although within a co-continuous structure with its thermoplastic counterpart, is transformed into dispersed particles encapsulated by the thermoplastic polymer. This morphological transformation is widely known as phase inversion. The result is a material most likely with rubber-like properties. Furthermore, dynamically vulcanized blends are melt-processable through existing thermoplastic processing equipments and their final morphology is stable and cannot be altered by any downstream operation. These special advantages of dynamically vulcanized blends have led the interest of both academia and industry toward the improvement of properties through continuously optimizing the parameters affecting their morphology development.

It is known that the elastomer phase in the initial stage of vulcanization is strongly deformed into continuous elastomeric threads and eventually breaks up and forms the final dispersed cross-linked domains [58]. As mentioned earlier, the presence of an initial co-continuous morphology prior to dynamic vulcanization is a prerequisite in obtaining fine dispersed cross-linked elastomer phase at the end of vulcanization [58,59]. From a morphological point of view, only an initial co-continuous morphology results in an effective and overall transfer of the shear and elongation stresses from one phase to the other and, hence, guarantees the aforementioned breakup of the elastomeric component [58]. The effectiveness of stress transfer and morphology transformation in an initially co-continuous morphology can be visualized by considering the complete opposite hypothetical situation, i.e., an initial dispersed/matrix morphology with the elastomeric component as the droplet phase. In this system, the viscosity and elasticity of the elastomeric component, which already forms the dispersed phase, increase during dynamic vulcanization. As a result, the stress transferred by the low viscosity thermoplastic phase becomes less and less effective in deforming the elastomeric domains and coalescence of the elastomeric domains becomes more and more hindered due to increased viscosity and elasticity of this phase. Consequently, the cross-linking reaction merely stabilizes the already existing, rather coarse dispersed morphology of the elastomeric domains, without any further morphological refinement [60-62]. With this in mind, although the co-continuity is a crucial factor in this process, a stable and unchangeable co-continuous morphology in the intermediate stage of dynamic vulcanization is not a desirable situation. This has been observed in ethylene methyl acrylate and linear low density polyethylene (EMA/LLDPE) blends where phase inversion was hindered due to the presence of a stable co-continuous morphology [63].

To obtain an optimum initial morphology prior to dynamic vulcanization, all the aforementioned parameters discussed in the previous section have to be taken into consideration. For instance, a blend with extremely low viscosity minor thermoplastic phase (low viscosity ratio, $\eta_{thermoplastic}/\eta_{elastomer}$) generally forms a dispersed/matrix morphology, where the thermoplastic phase encapsulates coarse elastomeric domains [61]. This is a condition which has to be avoided. On the opposite situation, a highly viscous thermoplastic phase (a high viscosity ratio blend) hinders the dispersion of the cross-linked elastomer particles [62]. The fact that high and low viscosity ratio systems represent completely different initial morphological states prior to dynamic vulcanization means that the torque requirement for mixing could be substantially different from one system to another. Indeed, blends with high initial viscosity ratio with a dispersed thermoplastic phase in an elastomeric matrix have demonstrated a shoulder in the mixing torque during dynamic vulcanization process [61]. This shoulder appears while the blend structure passes through a co-continuous morphology prior to complete vulcanization and dispersion of the elastomeric component in the thermoplastic phase. The shoulder was seen to disappear with less overall mixing torque requirement when the initial morphological state was already a co-continuous type. The appearance of a shoulder in mixing curves has been reported several times and it has been attributed to the onset of phase inversion [60,64]. From an industrial point of view mainly based on the energy consumption during the dynamic vulcanization step, an initial co-continuous structure is desired over the dispersed/matrix one. The co-continuous state guarantees a smoother phase transition with lower energy consumption (lower torque requirement).

In several cases, however, the appearance of a shoulder in the mixing torque could be overshadowed and the phase inversion process could not be easily detected due to the rapid cross-linking reaction [65]. Hence, the kinetics of cross-linking can also play a major role in the phase morphology development during dynamic vulcanization. Generally, the rate of the cross-linking reaction increases with temperature, especially in a blend with a heat reactive curing system such as phenolic resin [14,64,66]. The higher the temperature, the shorter is the time required to reach a certain level of cross-linking or gel content. The consequence of a faster reaction has been associated with the hindrance in the complete disintegration and dispersion of cross-linked elastomer phase during dynamic vulcanization [60]. At comparatively slower reaction rates, a smoother phase inversion with a better dispersion of elastomeric domains could be usually expected. With similar idea in mind, it has already been observed that the dispersion of a pre-cross-linked elastomer in a thermoplastic phase is a serious challenge especially when the gel content of the elastomer is larger than 70% [67]. Some other reactive systems other than thermoplastic/elastomer blends are as well shown to be affected by the effect of excessive gel content when dispersing one phase into the other. The dispersion of an *in situ* cured epoxy into a polystyrene (PS) is an example, where coalesced and agglomerated structure has been observed for gel content larger than 70% [68].

Beside the kinetics of the curing reaction, the mixing intensity is also a crucial factor which clearly affects the morphology of reactive blends. By mixing intensity, we mainly consider the apparent shear rate of the mixing system. Although this is an important mixing parameter, it cannot be easily dissociated from the effect of temperature through viscous dissipation. In a simple argument, one may observe an increase in the mixing temperature by increasing the rotational speed of the mixing equipment, i.e., by increasing the intensity of mixing. This additional increase in temperature could readily affect the morphology development through its influence on the rate of the cross-linking reaction. In a situation like that, the morphology development is said to be mainly controlled by the fragmentation of the cross-linked elastomer rather than by the transient equilibrium between coalescence and breakup of particles [61]. It has been pointed out that the significant effect of mixing intensity on the final morphology is through its effect on the rate of the cross-linking reaction [60,63].

Mixing intensity can be otherwise investigated by the use of different types of mixing equipment, e.g., internal batch mixers or twin-screw extruders. Regardless of the mixing equipment, the transition between co-continuous to dispersed-type morphologies with well-dispersed cross-linked elastomer domains has merely been attributed to the state of gel content in the blend [64]. This basically means that for a phase transition to happen the gel content of the mixture has to reach a certain level and, if an appropriate blend formulation with well-designed mixing conditions is set in advance, the transition will eventually happen at a similar point in the reaction regardless of the mixing equipment. In most cases with rapid reactive curing systems, the phase transformation occurs quite fast within the first one minute upon the addition of the curing system [58]. Despite this rapid onset, the final morphology of dynamically vulcanized blends depends on both the rate of the curing reaction and the total amount of shear exerted on the blend [69]. Large and to some extent coarse and interconnected vulcanized elastomeric domains have been reported for those dynamically cross-linked blends obtained in a twin-screw extruder in comparison to the ones from an internal mixer [14,69]. The observed morphological features have been mainly attributed to the fast curing reaction (due to larger viscous dissipation) and a shorter residence time in an extrusion process. A morphology comparison for dynamically vulcanized EPDM/PP blends is shown in Fig. 6. The mixture blended in an internal mixer resulted in distinct EPDM domains. On the other hand, in a twin-screw extrusion process, coarse, ruptured, and to some extent large and interconnected domains of EPDM appeared at the end of process.

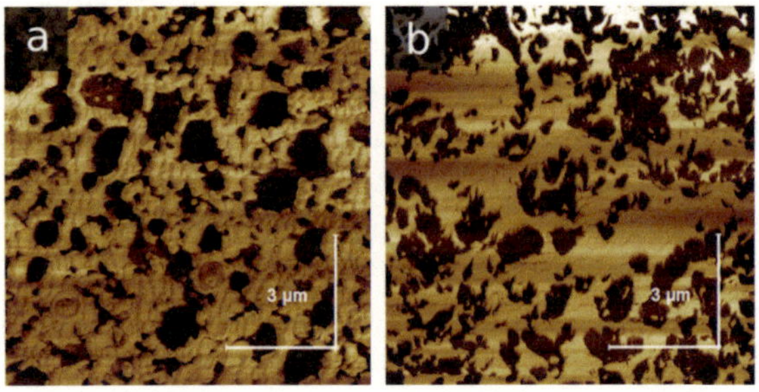

Figure 6. Atomic force microscopy images of reactive nonplasticized EPDM/PP 50/50 (wt/wt%): (a) internal mixer, (b) twin-screw extruder (Dark phase: EPDM; Bright phase: PP) (from [14])

The fast curing reaction in a twin-screw extrusion process in comparison to the reaction rate in an internal mixer has readily been quantified by the gel content of the final blend [14]. In blends with especially low elastomer content, the distinguishable effect of the mixing equipment on the reaction rate can be observed in Fig. 7. One may notice that at 25 and 40 wt% of EPDM, the final gel content of the elastomeric phase is significantly lower when dynamic vulcanization is performed in an internal mixer.

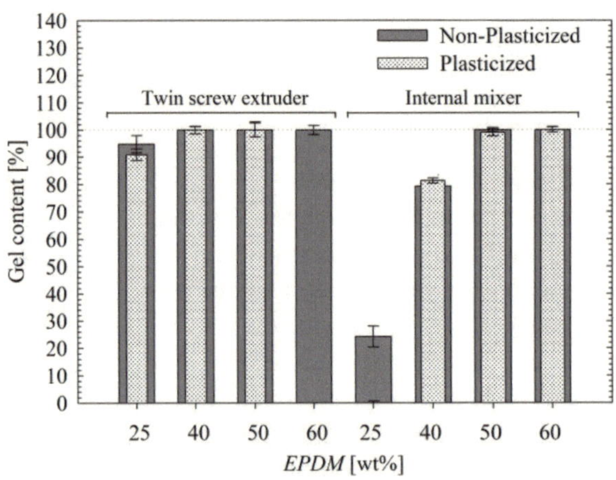

Figure 7. Gel content of the EPDM phase in both nonplasticized and plasticized dynamically vulcanized EPDM/PP blends (from [14])

In several reactive blends, other than the main polymeric components and the curing system, a plasticizer may also be incorporated. Already, the presence of a plasticizer is known to affect the morphological state prior to the dynamic vulcanization step. As it is discussed in the previous section, it is expected to observe phase morphology with larger and percolated elastomeric domains in plasticized blends prior to the addition of a curing system (Figs. 4 and 5). Meanwhile, whether a plasticizer could affect the rate of the cross-linking reaction may actually provide further information regarding the morphology evolution during dynamic vulcanization. Several characterization techniques such as gel content measurements and thermal analyses by differential scanning calorimetry are usually employed to evaluate a curing reaction. However, the rheological characterization of an elastomer compound by means of an oscillating disc rheometer (ODR) or a conventional rheometer is the most common technique to follow the curing behavior of an elastomer. In a conventional rheometer, one may measure the dynamic rheological properties such as the storage and loss moduli, i.e., G' and G'', and obtain the crossover point between these two quantities. The aforementioned crossover is the point where the elastic properties of the material overcome the viscous ones. An easy and convenient explanation is that after the crossover point during the curing reaction, the elastomer becomes more and more capable of storing the mechanical energy rather than dissipating it by means of internal friction. As a result, this rheological point has a special importance and it is attributed to the onset of network like behavior in polymeric systems. By returning to the subject of plasticization and its effect on the cross-linking reaction, the crossover point in plasticized elastomer has been clearly shown to be reached at longer times when compared to an identical nonplasticized elastomer [66]. As an example, the effects of both temperature and plasticization on the crossover point of an EPDM containing phenolic curing system are shown in Fig. 8.

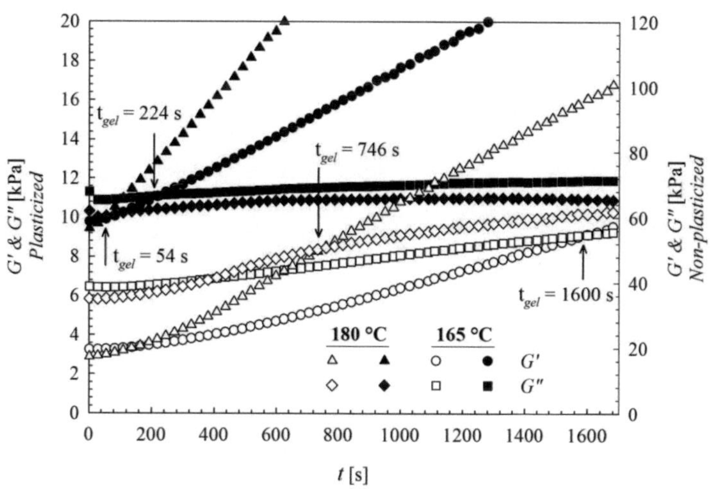

Figure 8. Effects of plasticization and temperature on the curing behavior of EPDM (filled symbols: nonplasticized; open symbols: plasticized; the crossovers are indicated by arrows) (from [66])

Performing such measurement is sometimes impossible for a reactive blend with a complete set of ingredients, i.e., both elastomer and thermoplastic phases in the presence of a curing system and plasticizer. The difficulty lies in the fact that in most cases it is impossible to inhibit the curing reaction while mixing with high-melting-point thermoplastic phase. If one needs to understand the effect of certain ingredients, such as plasticizer, on the curing behavior of complete formulation, two logical paths could be envisaged. The first one is to design a special formulation with no interference between the mixing conditions, e.g., temperature, and those which may initiate the curing process [66]. In a specially designed reactive EPDM/PP blend, the presence of a plasticizer was shown to induce an initial delayed reaction with a subsequent rapid curing behavior [66]. The second approach toward characterizing the effect of a plasticizer on the curing behavior is based on the gel content analysis of dynamically vulcanized blends [14]. A low level of gel content in the presence of a plasticizer at the end of dynamic vulcanization, especially for the low elastomer content reactive blends is already reported in Fig. 7. These observations, both on the neat elastomer and dynamically vulcanized formulations, indicate that the presence of a plasticizer induces a delayed reaction. How could this affect the morphology development in plasticized dynamically vulcanized blends? According to the previous discussion in this section, a slightly slower but gradually cross-linking reaction could actually result in smooth and good dispersion of the elastomer without its excessive rupture. However, a delayed reaction, especially in a flow field of low extensional and/or shear deformations, may actually result in large coalesced elastomer domains, which is an undesired morphological state in the case of dynamically vulcanized blends [66]. The formation of such coalesced structure in the presence of plasticizer can be observed in Fig. 9, where a plasticized reactive EPDM/PP was subjected to an *in situ* curing and shearing at 0.1 s^{-1} in a rheometer. One may see that the initial co-continuous morphology was gradually transformed into elongated elastomeric domains, where afterward at later stage and longer times (Fig. 9d), large domains of EPDM were formed mainly due to coalescence of elastomeric domains.

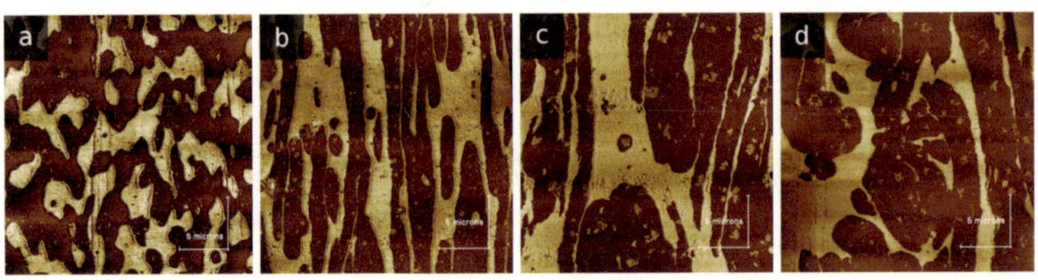

Figure 9. Atomic force microscopy images of a plasticized EPDM/PP 50/50 (wt/wt%) reactive blend at different shearing stages in a rheometer at 165 °C and 0.1 s^{-1}: (a) initial morphology prior shearing and curing, (b) after 450 s, (c) after 2700 s, (d) after 7200 s (Dark phase: EPDM; Bright phase: PP) (from [66])

In a similar nonplasticized reactive blend that was subjected to an identical flowing condition, a complete different morphology evolution could be observed (Fig. 10). First of all, the initial morphology is comparably finer. Furthermore, the gradual shearing with the no delayed curing reaction transforms the co-continuous morphology into a structure where less coalescence could be observed. More interestingly, at longer times (Fig. 10d) the thermoplastic phase shows the tendency of encapsulating the elastomeric domains. This morphological transformation is basically what is wished for when dynamic vulcanization process is in mind.

To conclude, the ideal morphological state at the end of the dynamic vulcanization process with finely dispersed elastomeric particles could only be achieved with the finest initial morphology and with a well-designed formulation and processing parameters. All these together, in a perfectly fine-tuned mixing procedure based on a correct blending sequence with

http://dx.doi.org/10.5772/61414

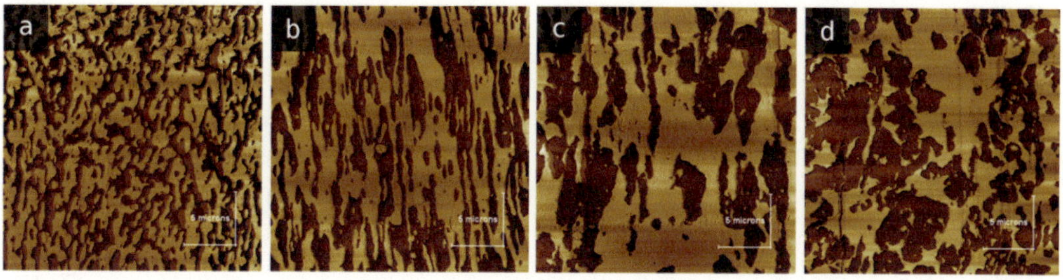

Figure 10. Atomic force microscopy images of a nonplasticized EPDM/PP 50/50 (wt/wt%) reactive blend at different shearing stages in a rheometer at 165 °C and 0.1 s^{-1}: (a) initial morphology prior shearing and curing, (b) after 450 s, (c) after 2700 s, (d) after 7200 s (Dark phase: EPDM; Bright phase: PP) (from [66])

respect to all ingredients and their corresponding effects on the phase morphology evolution, will guarantee the success of the dynamic vulcanization process.

5. Concluding remarks

The morphology evolution and the state of the final phase structure have crucial effects on the physical and mechanical properties of a reactive thermoplastic/elastomer system. A successful approach toward product development of such materials usually begins by identifying the essential requirements of the final product. Accordingly, the ingredients should be carefully chosen as each fulfills a special role in the overall formulation. Thermoplastic elastomers may contain several different chemicals such as polymeric components, curing system, fillers, additives, and plasticizer. Product development could only be successful if the role and the effect of each ingredient on the phase morphology are thoroughly investigated in advance. Throughout this procedure, the rheological, interfacial, and thermal properties of complete formulations with the curing behavior have to be all considered. Only based on this information, an effective mixing and blending strategy can be established, which at the end will guarantee the desired target properties.

Author details

Shant Shahbikian[1] and Pierre J. Carreau[2*]

*Address all correspondence to: pcarreau@polymtl.ca

1 CREPEC, Chemical and Biotechnological Engineering Department, University of Sherbrooke, QC, Canada

2 CREPEC, Chemical Engineering Department, Polytechnique Montreal, Montreal, QC, Canada

References

[1] J.G. Drobny, *Handbook of Thermoplastic Elastomers*, William Andrew Pub., Norwich, NY (2007), pp. 1-8.

[2] G. Holden, *Understanding Thermoplastic Elastomers*, Hanser Gardner Pub., Cincinnati, OH (2000), pp. 15-53, 65-74.

[3] G. Holden, H.R. Kricheldorf and R.P. Quirk, *Thermoplastic Elastomers*, Hanser, Cincinnati, OH (2004), pp. 109-117.

[4] S.K. De and J.R. White, *Rubber Technologist's Handbook*, Rapra Technology Limited (2001), pp. 107-112.

[5] A.M. Gessler and W.H. Haslett, Jr., US, 3,037,954 (1962).

[6] W.K. Fischer, US Patent 3,862,106 (1975).

[7] W.K. Fischer, US, 3,835,201 (1974).

[8] W.K. Fischer, US, 3,806,558 (1974).

[9] W.K. Fischer, US Patent 3,758,643 (1973).

[10] A.Y. Coran, B. Das and R.P. Patel, US Patent 4,130,535 (1978).

[11] A.Y. Coran and R.P. Patel, US Patent 4,104,210 (1978).

[12] A.Y. Coran and R.P. Patel, US Patent 4,130,534 (1978).

[13] S. Sabet-Abdou and M.A. Fath, US Patent 4,311,628 (1982).

[14] S. Shahbikian, P.J. Carreau, M.C. Heuzey, M.D. Ellul, J. Cheng, P. Shirodkar and H.P. Nadella. Morphology development of EPDM/PP uncross-linked/dynamically cross-linked blends. Polym Eng Sci. 2012; 52(2): 309-322.

[15] M.D. Ellul. Plasticization of polyolefin elastomers, semicrystalline plastics and blends crosslinked in situ during melt mixing. Rubber Chem Technol. 1998; 71(2): 244-276.

[16] G. Wypych, *Handbook of Plasticizers*, Toronto - New York (2004), pp. 193-380.

[17] A.K. Doolittle. Mechanism of solvent action. Ind Eng Chem. 1946; 38(5): 535-540.

[18] A.K. Doolittle. Mechanism of solvent action. Ind Eng Chem. 1944; 36(3): 239-244.

[19] R. Houwink, *Proceedings of the XIth International Congress of Pure and Applied Chemistry*, 575-583 (1947).

[20] F.W. Clark. Plasticizers. Soc of Chem Ind – J Chem Ind. 1941; 60(14): 225-228.

[21] A. Kirkpatrick. Some relations between molecular structure and plastricizing effect. J Appl Phys. 1940; 11(4): 255-267.

[22] J.T.G. Fox and P.J. Flory. Second-order transition temperatures and related properties of polystyrene -- I. Influence of molecular weight. J Appl Phys. 1950; 21(6): 581-591.

[23] J.K. Sears and J.R. Darby, *The Technology of Plasticizers*, Wiley, New York, NY (1982), pp. 35-77.

[24] P.J. Flory. Thermodynamics of High Polymer Solutions. J Chem Phys. 1941; 9(8): 660.

[25] M.L. Huggins. Solutions of Long Chain Compounds. J Chem Phys. 1941; 9(5): 440.

[26] J. Brandrup, E.H. Immergut and E. Grulke, *Polymer Handbook*, Interscience Publishers (1999).

[27] J.E. Mark, *Polymer Data Handbook*, Oxford University Press (2009).

[28] J.H. Hildebrand and R.L. Scott, *The Solubility of Nonelectrolytes*, Dover Publications, Inc, New York, NY (1964), pp. 119-133.

[29] D. Mangaraj. Elastomer blends. Rubber Chem Technol. 2002; 75(3): 365-427.

[30] C.M. Hansen, *Hansen Solubility Parameters: A User's Handbook*, CRC Press, Boca Raton (2007), pp. 1-26.

[31] S. Abbott, Chemical Compatibility of Poly(lactic acid): A Practical Framework Using Hansen Solubility Parameters, Auras, R., In: *Poly(lactic acid): Synthesis, Structures, Properties, Processing, and Applications*, Wiley: Hoboken, NJ (2010), pp. 83-95.

[32] V. Mishra, D.A. Thomas and L.H. Sperling. Partition coefficient for small amounts of monomer, solvent, or plasticizer in a two-phased polymer blend or IPN. J Polym Sci Pol Phys. 1996; 34(12): 2105-2108.

[33] H. Xinping and S. Kok Siong. Mechanical properties and ionic conductivities of plasticized polymer electrolytes based on ABS/PMMA blends. Polymer. 2000; 41(24): 8689-8696.

[34] B. Ohlsson, H. Hassander and B. Tornell. Blends and thermoplastic interpenetrating polymer networks of polypropylene and polystyrene-block-poly(ethylene-stat-butylene)-block-polystyrene triblock copolymer.1. Morphology and structure-related properties. Polym Eng Sci. 1996; 36(4): 501-510.

[35] K. Jayaraman, V.G. Kolli, S.-Y. Kang, S. Kumar and M.D. Ellul. Shear flow behavior and oil distribution between phases in thermoplastic vulcanizates. J Appl Polym Sci. 2004; 93(1): 113-121.

[36] W.G.F. Sengers, P. Sengupta, J.W.M. Noordermeer, S.J. Picken and A.D. Gotsis. Linear viscoelastic properties of olefinic thermoplastic elastomer blends: melt state properties. Polymer. 2004; 45(26): 8881-8891.

[37] W.G.F. Sengers, M. Wubbenhorst, S.J. Picken and A.D. Gotsis. Distribution of oil in olefinic thermoplastic elastomer blends. Polymer. 2005; 46(17): 6391-6401.

[38] R. Winters, J. Lugtenburg, V.M. Litvinov, M. van Duin and H.J.M. de Groot. Solid state C-13 NMR spectroscopy on EPDM/PP/oil based thermoplastic vulcanizates in the melt. Polymer. 2001; 42(24): 9745-9752.

[39] T. Abraham, N.G. Barber and M.P. Mallamaci. Oil distribution in iPP/EPDM thermoplastic vulcanizates. Rubber Chem Technol. 2007; 80(2): 324-339.

[40] C.E. Scott and C.W. Macosko. Model experiments concerning morphology development during the initial stages of polymer blending. Polym Bull. 1991; 26(3): 341-348.

[41] C.K. Shih, D.G. Tynan and D.A. Denelsbeck. Rheological properties of multicomponent polymer systems undergoing melting or softening during compounding. Polym Eng Sci. 1991; 31(23): 1670-1673.

[42] R. Ratnagiri and C.E. Scott. Phase inversion during compounding with a low melting major component: Polycaprolactone/polyethylene blends. Polym Eng Sci. 1998; 38(10): 1751-1762.

[43] B.D. Favis and J.-P. Chalifoux. Effect of viscosity ratio on the morphology of polypropylene/polycarbonate blends during processing. Polym Eng Sci. 1987; 27(21): 1591-1600.

[44] B.D. Favis and J.-P. Chalifoux. Influence of composition on the morphology of polypropylene/polycarbonate blends. Polymer. 1988; 29(10): 1761-1767.

[45] J. Karger-Kocsis, A. Kallo and V.N. Kuleznev. Phase structure of impact-modified polypropylene blends. Polymer. 1984; 25(2): 279-286.

[46] S. Wu. Formation of dispersed phase in incompatible polymer blends: interfacial and rheological effects. Polym Eng Sci. 1987; 27(5): 335-343.

[47] K.A. Moly, Z. Oommen, S.S. Bhagawan, G. Groeninckx and S. Thomas. Melt rheology and morphology of LLDPE/EVA blends: Effect of blend ratio, compatibilization, and dynamic crosslinking. J Appl Polym Sci. 2002; 86(13): 3210-3225.

[48] R.C. Willemse, A. Posthuma De Boer, J. Van Dam and A.D. Gotsis. Co-continuous morphologies in polymer blends: The influence of the interfacial tension. Polymer. 1999; 40(4): 827-834.

[49] S. Shahbikian, P.J. Carreau, M.C. Heuzey, M.D. Ellul, H.P. Nadella, J. Cheng and P. Shirodkar. Rheology/Morphology Relationship of Plasticized and Nonplasticized Thermoplastic Elastomers Based on Ethylene-Propylene-Diene-Terpolymer and Polypropylene. Polym Eng Sci. 2011; 51(11): 2314-2327.

[50] W. Bu and J. He. Effect of mixing time on the morphology of immiscible polymer blends. J Appl Polym Sci. 1996; 62(9): 1445-1456.

[51] R.C. Willemse, A.P. De Boer, J. Van Dam and A.D. Gotsis. Co-continuous morphologies in polymer blends: a new model. Polymer. 1998; 39(24): 5879-5887.

[52] J.M. Li, P.L. Ma and B.D. Favis. The role of the blend interface type on morphology in cocontinuous polymer blends. Macromolecules. 2002; 35(6): 2005-2016.

[53] B.D. Favis. Effect of processing parameters on the morphology of an immiscible binary blend. J Appl Polym Sci. 1990; 39(2): 285-300.

[54] U. Sundararaj, C.W. Macosko, R.J. Rolando and H.T. Chan. Morphology development in polymer blends. Polym Eng Sci. 1992; 32(24): 1814-1823.

[55] P.G. Ghodgaonkar and U. Sundararaj. Prediction of dispersed phase drop diameter in polymer blends: The effect of elasticity. Polymer Engineering and Science. 1996; 36(12): 1656-1665.

[56] F. Goharpey, R. Foudazi, H. Nazockdast and A.A. Katbab. Determination of twin-screw extruder operational conditions for the preparation of thermoplastic vulcanizates on the basis of batch-mixer results. J Appl Polym Sci. 2008; 107(6): 3840-3847.

[57] S. Thomas and G. Groeninckx. Nylon 6/ethylene propylene rubber (EPM) blends: Phase morphology development during processing and comparison with literature data. J Appl Polym Sci. 1999; 71(9): 1405-1429.

[58] H.-J. Radusch, Phase Morphology of Dynamically Vulcanized Thermoplastic Vulcanizates, Harrats, C.; Thomas, S.; Groeninckx, G., In: *Micro- and Nanostructured Polymer Blend Systems: Phase Morphology and Interfaces*, CRC/Taylor & Francis: Boca Raton, FL (2006).

[59] K. Naskar and J.W.M. Noordermeer. Thermoplastic elastomers by dynamic vulcanization. Prog Rubber Plast Recyc Technol. 2005; 21(1): 1-26.

[60] N. Dufaure, P.J. Carreau, M.C. Heuzey and A. Michel. Phase inversion in immiscible blends of PE and reactive EVA. J Polym Eng. 2005; 25(3): 187-216.

[61] C. Joubert, P. Cassagnau, A. Michel and L. Choplin. Influence of the processing conditions on a two-phase reactive blend system: EVA/PP thermoplastic vulcanizate. Polym Eng Sci. 2002; 42(11): 2222-2233.

[62] J. Oderkerk and G. Groeninckx. Morphology development by reactive compatibilisation and dynamic vulcanisation of nylon6/EPDM blends with a high rubber fraction. Polymer. 2002; 43(8): 2219-2228.

[63] A. Bouilloux, B. Ernst, A. Lobbrecht and R. Muller. Rheological and morphological study of the phase inversion in reactive polymer blends. Polymer. 1997; 38(19): 4775-4783.

[64] A. Verbois, P. Cassagnau, A. Michel, J. Guillet and C. Raveyre. New thermoplastic vulcanizate, composed of polypropylene and ethylene-vinyl acetate copolymer crosslinked by tetrapropoxysilane: evolution of the blend morphology with respect to the crosslinking reaction conversion. Polym Int. 2004; 53(5): 523-535.

[65] F. Goharpey, A.A. Katbab and H. Nazockdast. Mechanism of morphology development in dynamically cured EPDM/PP TPEs. I. Effects of state of cure. J Appl Polym Sci. 2001; 81(10): 2531-2544.

[66] S. Shahbikian, P.J. Carreau, M.C. Heuzey, M.D. Ellul, H.P. Nadella, J. Cheng and P. Shirodkar. Morphology and rheology of nonreactive and reactive EPDM/PP blends in transient shear flow: Plasticized versus nonplasticized blends. Rubber Chem Technol. 2011; 84(3): 325-353.

[67] G. Martin, C. Barres, P. Sonntag, N. Garois and P. Cassagnau. Morphology development in thermoplastic vulcanizates (TPV): Dispersion mechanisms of a pre-crosslinked EPDM phase. Eur Polym J. 2009; 45(11): 3257-3268.

[68] F. Fenouillot and H. Perier-Camby. Formation of a fibrillar morphology of crosslinked epoxy in a polystyrene continuous phase by reactive extrusion. Polym Eng Sci. 2004; 44(4): 625-637.

[69] P. Sengupta and J.W.M. Noordermeer. Effects of composition and processing conditions on morphology and properties of thermoplastic elastomer blends of SEBS-PP-oil and dynamically vulcanized EPDM-PP-oil. J Elastom Plast. 2004; 36(4): 307-331.

Chapter 4

Environmental Degradability of Polyurethanes

Katarzyna Krasowska, Aleksandra Heimowska and Maria Rutkowska

Additional information is available at the end of the chapter

http://dx.doi.org/10.5772/60925

Abstract

The growing interest in environmental issues and increasing demands to develop materials that do not burden the natural environment significantly are currently observed. In this connection many studies on polymer degradation in different environments are carried out. It is important to consider the influence of synergistic action of various factors in order to understand the environmental degradation of synthetic polymers. This requires understanding of interactions between polymer and living organisms.

This paper reviews current authors research on environmental degradation of polyurethanes.

The comparison of environmental degradability of polyurethanes in the Baltic Sea water and compost under natural weather depending conditions is presented. The environmental degradation of poly(ester-urethane) based on poly(ethylene-butylene-adipate) and poly(ester-urethane) based on poly (ε -caprolactone) was evaluated.

The characteristic parameters of sea water (temperature, pH, salt, and oxygen contents) and of compost (temperature, pH, moisture content, and activity of dehydrogenases) were monitored and their influence on degradation of polyurethanes was discussed.

The environmental degradability of polyurethanes was investigated by changes of weight, tensile strength, morphology, and crystallinity of polyurethanes after incubation in environment. The investigated polyurethanes were degradable in both natural environments and their environmental degradability depends on the chemical structure and the kind and conditions of environment.

Keywords: Poly(ester-urethane), environmental degradation, sea water, compost

1. Introduction

Polyurethanes are an important and versatile class of synthetic polymers that can be synthesized on a large scale and can be processed in a variety of different ways. For these reasons, polyurethanes are widely used in many aspects of modern life. They have found a widespread use in the industrial fields, for example as furniture coatings, adhesives, construction materials, flame retardants, fibers, paddings, paints, elastomeric parts, and synthetic skins [1-5]. Moreover, over the past 40 years polyurethanes have also been used in various biomedical applications such as vascular prostheses, artificial skin, pericardial patches, soft-tissue adhesive, drug delivery devices, and scaffolds for tissue engineering [6-15].

Polyurethane is a general term used for a class of polymers that are synthesized from three basic components: a diisocyanate, a polyol, and a chain extender. The terminal hydroxyl group allows for alternating blocks, called "segments", to be inserted into polyurethane chain. Soft segments are derived from polyols such as polyester, while the hard segments are formed from the combination of diisocyanate and a chain extender component. The chain extender is usually a small molecule with either hydroxyl or amine groups.

On the one hand, hard segments contribute to hardness, tensile strength, impact resistance, stiffness, and modulus. On the other hand, soft segments contribute to water absorption, elongation, elasticity, softness, and degradability. Hence, from the point of view of applications, it is possible to produce various polyurethanes, which properties can be easily modified by varying structures of soft and hard segments [16].

The nature of polyurethanes chemistry is the central point for understanding why some polyurethanes are non-degradable and other undergo fast degradation. Both non-degradable and degradable polyurethanes can be designed through a proper selection of building segments. Non-degradable polyurethanes are characterized by their excellent chemical stability, abrasion resistance, and mechanical properties. Now despite the xenobiotic origins of polyurethanes, they have been found to be susceptible to degradation by naturally occurring microorganisms.

Degradable polyurethanes are generally achieved by incorporating labile and hydrolysable moieties into polymer chain. It is well known that the biodegradation of polyurethanes depends on their structure, which is conditioned by several factors such as the nature of the used polyol. It has been reported that poly(ester-urethanes) are prone to microbial degradation due to the presence of ester bonds that are known to be enzymatic hydrolysable.

Incorporation of biodegradable fillers or biodegradable aliphatic isocyanates can also enhance biodegradability of polyurethanes and then reduce negative influence on the environment. For example, using a lignin-based polyols in the polyurethanes was observed higher fungal degradation level [17].

Considering the variety of possible applications very often the degradability of polyurethanes has become an important or even a deciding factor. For example, polyurethanes used for

http://dx.doi.org/10.5772/60925

insulation of under-water cables should have high resistance to environmental degradation. In contrast to this, biodegradable polyurethane wastes could be place for utilisation or accumulation in natural environment.

Degradation of polymers is determined by different factors; apart from polymer type the nature of the environment is also important.

The initial breakdown of a polymer, which is the first step of the biological degradation process, can result from physical and biological forces. Physical forces such as heating/cooling, freezing/thawing, or wetting/drying can cause mechanical damage such as cracking of polymeric materials. The growth of many microorganisms can also cause small-scale swelling and bursting of polymeric materials. Most polymers are too large to pass through cellular membranes, so they must be depolymerized to smaller monomers before they can be adsorbed and degraded within microbial cells. The monomers, dimers, and oligomers of a polymer's repeating units are much easily degraded and mineralized because they can be assimilated through the cellular membrane and then further degraded by cellular enzymes. Two categories of enzymes are involved in the biological degradation of polymers: extracellular and intracellular depolymerases. During degradation, exoenzymes from microorganisms break down complex polymers, yielding smaller molecules of short chains that are small enough to pass semi-permeable outer bacterial membranes and then to be utilized as carbon energy sources. Under oxygen conditions, aerobic microorganisms are mostly responsible for the degradation of polymer. Biomass, carbon dioxide, and water are the final products of deterioration. As opposite to this, under anoxic conditions, anaerobic microorganisms play the main role in polymer destruction. The primary products are methane, water, and biomass [18].

Figure 1 represents the general mechanism of polymer biological degradation under aerobic conditions in a natural environment such as sea water and compost, which are studied by authors.

According to the literature microorganisms such as fungi and bacteria are involved in the degradation process of polyurethanes [18].

Generally, three types of polyurethane degradations have been identified: fungal degradation, bacterial degradation, and degradation by polyurethanase enzymes. However, polyurethanes are especially susceptible to fungal attack. Soil fungal communities are involved in polyurethanes degradation. For example, four species of fungi, namely, *Curvularia senegalensis, Fusrium solani, Aureobasidium pullulans and Cladosporium,* were obtained from soil and found to degrade ester-based polyurethanes [2, 19]. Bacteria known to degrade poly(ester-urethanes) also produce polyurethanes degrading enzymes, such as polyurethane-esterases. Two kinds of polyurethane-esterase enzymes, such as an intracellular polyurethane-esterase and an extracellular polyurethane-esterase, play predominant and various roles in polyurethane biodegradation process. The intracellular enzyme provides access to the hydrophobic polyurethane surface. Then the extracellular enzyme sticks on the surface of the polyurethane. During these enzymatic actions, the ability of bacteria to adhere to the polyurethanes surface and to hydrolyse polyurethane substrates to metabolites is observed [18].

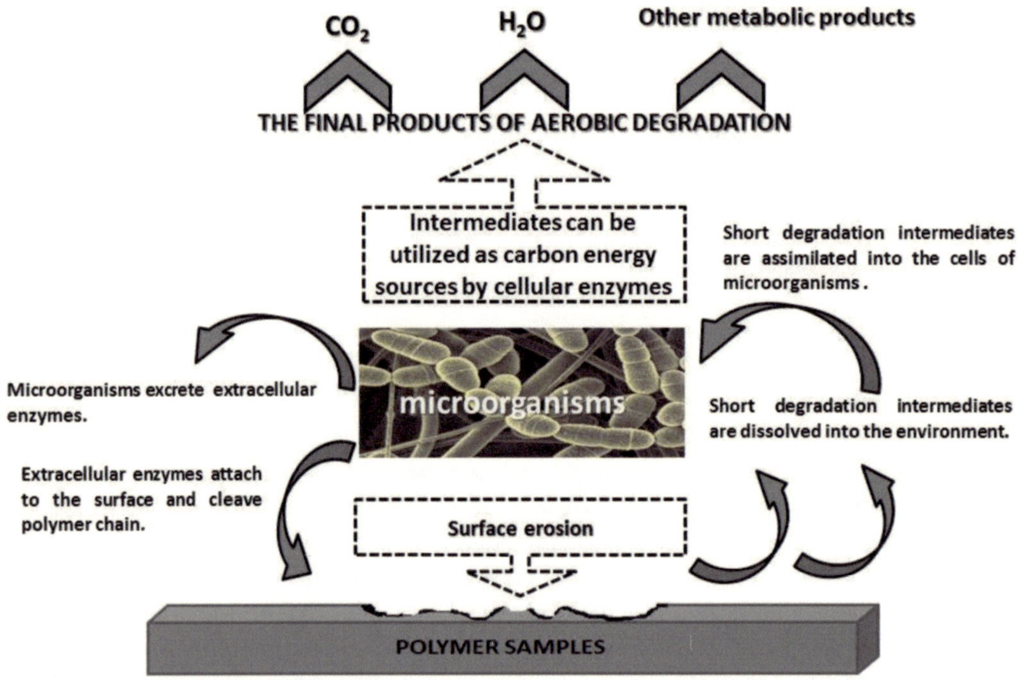

Figure 1. General mechanism of polymer biodegradation under aerobic conditions.

According to the literature, the degradation of polyurethanes is mostly studied in laboratories, in many cases under stable and favourable temperature conditions, providing additional nutrients to microorganisms and using highly concentrated enzymes to promote degradation [2, 4, 6, 8, 17, 20-24].

In the case of degradation in natural environment, very often the synergistic action of various factors leads to polymers degradation. Each natural environment contains different macroorganisms, microorganisms, and enzymes (in terms of species diversity and population). Different physical and chemical parameters, which have influence on rates of microbial activity, affect the rate of the degradation process. Thus, it is very interesting to estimate degradability of polyurethanes under natural weather - depending conditions.

The sea is a very complicated natural environment for degradation because microorganisms, animals, salt, sunlight, fluctuation of water, rain, etc., all play a part in degradation in nature. But it is known that a wide population of living organisms can also exist in the compost. Therefore, sea water and compost under natural weather depending conditions could be used for accumulation and utilization of polyurethane wastes.

The aim of this paper is to summarize and compare the results of environmental degradation of poly(ester-urethanes) in the Baltic Sea water [25] and in compost under natural weather depending conditions [26].

http://dx.doi.org/10.5772/60925

2. Experimental

2.1. Materials

Two kinds of poly(ester-urethanes) were used in this work: poly(ester-urethane)A and poly(ester-urethane)B designated as PU-A and PU-B respectively. The polyurethanes were obtained by a two-step condensation reaction [27, 28].

In the first step, the prepolymers were prepared from 4,4'diphenylmethane diisocyanate /MDI/ and different polydiols. The poly(ethylene-buthylene-adipate) /PEBA/, MW=2000 was used for poly(ester-urethane)A. Poly(caprolactone)diol /PCLD/, MW=2000 was used for poly(ester-urethane)B. The molar ratio of NCO:OH was 4:1 in all cases in the first step reaction. The synthesis was carried out at 80°C. Both prepolymers were further extended by 1,4-butanediol / BD/. A chain extender was added to the prepolymer in an appropriate quantity to maintain a steady NCO:OH ratio of 1.1:1. The content of hard segments (the reaction product of MDI and low molecular weight chain extender) in obtained PU-A and PU-B was 38% [25, 26].

After synthesis, the polymers obtained in a form of a sheet (thickness of 2 mm), were cut into dumbbell-shaped samples and were used for the incubation in the environments.

Before incubation in the environments, the swelling test of polyurethane samples [25] was carried out in acetone and tetrahydrofuran (THF) to estimate the partial solubility and swelling degree, which can inform us about the extent of crosslinking.

Looking at the results of swelling degree and mass loss presented in Table 1, we can notice that PU-A is soluble in THF, PU-B is partially soluble. Thus, both the studied poly(ester-urethanes) differ not only in their chemical structure but in their network as well. PU-B can be partially crosslinked by allophanate bonds while PU-A is uncrosslinked.

Polymer	Equilibrium swelling degree (q)*		Mass loss [%]	
Acetone	THF	Acetone	THF	
PU-A	2.2	Soluble	40	Soluble
PU-B	2.0	6.1	8.0	80

(*) Was calculated from the mass-ratio before and after swelling.

Table 1. Results of swelling test for the blind samples of polyurethanes in acetone and tetrahydrofuran.

2.2. Degradation environments

The incubation of poly(ester-urethanes) took place in two natural environments:

- the Baltic Sea water in Gdynia Harbour
- compost pile under natural weather depending conditions.

The environmental degradation of PU-A and PU-B was carried out for up to 24 months.

During the environmental degradation in the Baltic Sea water, the samples were located in a special basket at 2 m depth under the surface of the sea, near the ship of the Polish Ship Salvage Company. The place of environmental degradation of polyurethane samples in the harbour area of the Baltic Sea is shown on Figure 2.

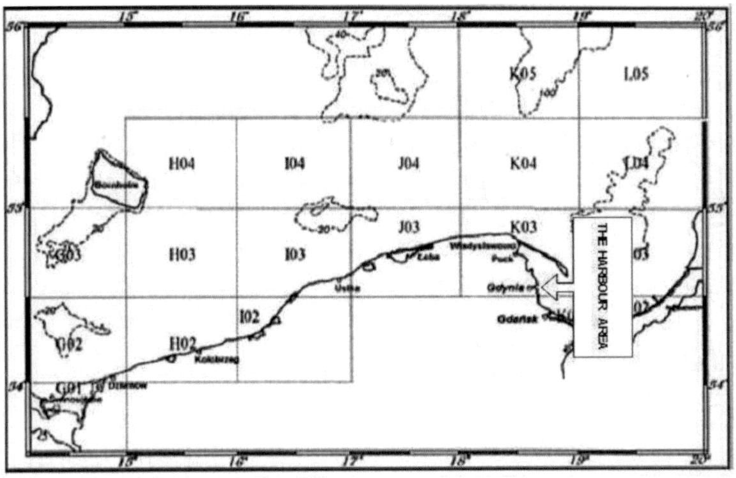

Figure 2. Place of environmental degradation of polyurethane samples in the Baltic Sea.

Degradation of polyurethane samples was also performed in the laboratory in a liquid medium containing sea water with NaN_3 (0.195 g NaN_3/1000cm^3). The polyurethane samples were located in the glass aquarium, with dimensions of 40 × 40 × 20 cm, equipped with an aeration pump. The sodium azide was added to the sea water to exclude the activity of microorganisms and to evaluate the resistance of the polymers to hydrolysis.

The characteristic parameters of sea water according to the Gdynia Water Management and Meteorology Institute and of liquid medium containing sea water with NaN_3 are shown in Table 2.

	PARAMETERS					
	Baltic Sea water				**sea water with NaN_3**	
Months	Temperature [°C]	pH	Oxygen contents [cm^3/dm^3]	Salt contents [ppt]	Temperature [°C]	pH
January	2.5	8.2	9.7	6.9	18.0	8.1
February	1.6	8.2	10.3	6.4	18.5	8.0
March	3.5	8.2	10.3	6.5	19.0	8.0
April	5.1	8.4	10.0	6.8	20.0	8.0
May	13.1	8.5	8.5	6.2	21.5	8.0
June	16.5	8.3	8,0	6.3	23.0	8.0

http://dx.doi.org/10.5772/60925

Months	PARAMETERS					
	Baltic Sea water				sea water with NaN_3	
	Temperature [°C]	pH	Oxygen contents [cm^3/dm^3]	Salt contents [ppt]	Temperature [°C]	pH
July	19.4	8.1	7.6	6.1	22.0	8.1
August	20.1	8.2	6.5	6.1	23.0	8.0
September	16.2	8.2	6.7	6.5	20.0	8.1
October	13.1	8.1	6.7	7.1	19.0	8.1
November	6.7	8.2	7.8	6.7	18.0	8.1
December	3.4	8.2	8.0	6.7	18.0	8.0

Table 2. The characteristic parameters of the Baltic Sea water and sea water with NaN_3.

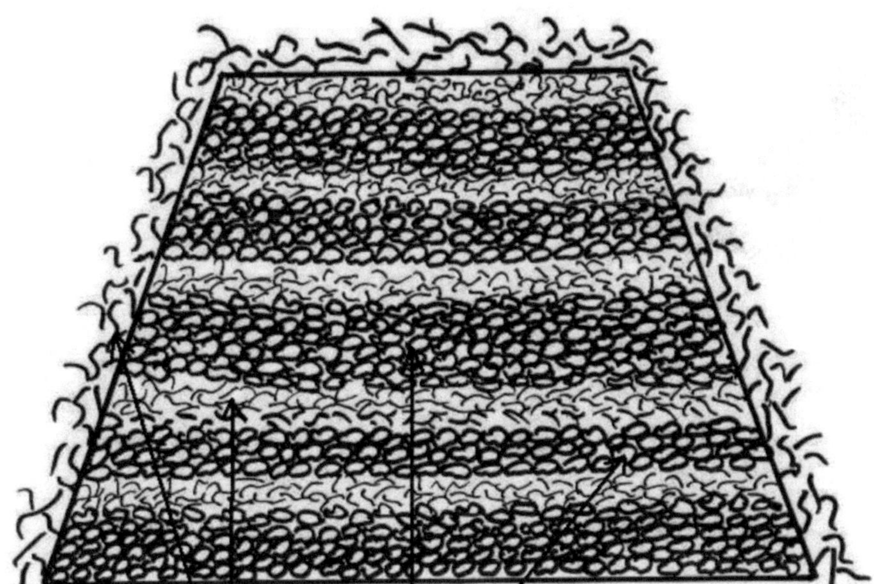

Figure 3. The cross-section of the compost pile prepared in natural environment [26].

The compost used in this work was formed with the dehydrated sewage sludge taken from a municipal waste treatment plant in Gdynia. The compost pile was prepared under natural conditions of sewage farm. It consisted of the sewage sludge, burnt lime, and straw. Burnt lime (0.45 kgCaO/1kg dry mass of compost) was added to ravage phatogenic bacterium and eggs parasites, to deacidificate sewage sludge, and to convert sludge to compost. The straw was added to maintain the higher temperature of the compost pile and to loosen the structure of the compost pile. The compost pile prepared under natural conditions was not adequately aerated, so it is expected that a combination of conditions from aerobic at the upper part of pile, microaerophilic in the middle part, and facultative anaerobic at the bottom of the pile could occur for microorganism growth [26]. Figure 3 represents cross-section of the compost pile in natural environment.

The characteristic parameters of the compost, such as, temperature, pH, moisture content, and activity of dehydrogenases, were measured during the degradation process and are shown in Table 3.

Months	PARAMETERS			
	Temperature [°C]	pH	Moisture content [%]	Activity of dehydrogenases [mol mg^{-1} d.m.]
January	6.0	5.3	49	0.0281
February	4.0	5.6	50	0.0286
May	15.0	5.7	55	0.0297
July	19.0	5.5	53	0.0328
August	22.0	5.4	56	0.0431
November	8.0	5.8	60	0.0318
December	7.0	5.9	55	0.0192

Table 3. The characteristic parameters of compost.

2.3. Measurements

2.3.1. Characterization of the compost pile under natural conditions

The characteristic parameters of the compost such as temperature, pH, moisture content, and activity of dehydrogenases were measured during degradation process.

2.3.1.1. The humidity of compost

The moisture content of the compost was determined by drying at 105°C until constant weight was obtained.

2.3.1.2. The pH of the compost

The pH of the compost was determined with a Teleko N 5172 pH-meter [26].

2.3.1.3. The biochemical activity of compost

To estimate the biochemical activity of microorganisms in sludge, the activity of the dehydrogenases was measured by a spectrophotometric method using triphenyltetrazolium chloride (TTC). The method is based on the dehydrogenation of glucose added to the compost with a subsequent transfer of hydrogen to the colourless, biologically active compound of TTC, which undergoes a reduction to triphenyloformazan (TF). The intensity of red colour compound TF was measured using a Specol colorimeter at 490 nm [26].

2.3.2. Investigations of polyurethanes samples

After incubation time, the samples were taken out from the environment, washed with distilled water, and dried at room temperature until constant weight.

The environmental degradability of polyurethanes was investigated by changes of weight, tensile strength, morphology, and crystallinity of polyurethane samples after incubation in the environment.

2.3.2.1. The changes in the polymer surface

The changes of polyurethanes surface were evaluated at macro- and micro scale. The views of polyurethane samples surface before and after degradation were compared. The pictures were taken before and after incubation in the environment. Microscopic observations of samples surface were done at magnification of 1:300 using the optical microscope ALPHAPHOT-2YS2-H Nikon linked to the photo camera Casio QV-2900UX. The samples were observed with and without polarizer.

2.3.2.2. The changes of weight

Weight changes were determined using an electronic balance Gibertini E 42s. The weight of clean and dried samples of polyurethanes after incubation in the compost was compared with those before incubation. The percentage weight changes [%] were calculated from the weight data.

2.3.2.3. The changes of tensile strength

Tensile strength [MPa] was measured at room temperature using a Tensile Testing Machine ZMGi-250 according to PN-EN ISO Standard [29].

2.3.2.4. The changes of thermal properties

Thermal analysis was carried out using Perkin-Elmer Pyris 1 Differential Scanning Calorimeter equipped with Intracooler 2P. The heating scans at the rate of 5 K/min in the temperature range -65-230°C in nitrogen atmosphere were recorded [26].

3. Results and discussion

3.1. Characteristics of the degradation environments of polyurethanes samples

The characteristic parameters of the Baltic Sea water and compost pile under natural conditions were monitored during the environmental degradation process of polyurethanes and their influence on degradation of polyurethanes was discussed. As both biotic and abiotic parameters of sea water (temperature, pH, salinity, and oxygen content) and compost (temperature, pH, moisture content, and activity of dehydrogenases) have a significant influence on the development of living microorganisms in natural environment.

Looking at the parameters presented in Tables 2 and 3, we can state, that the average temperature in the Baltic Sea was about 10°C and about 12°C in the compost pile. The temperature of both environments, depending on the weather conditions (season), had been fluctuating a lot during the experiment (from 1 to 20°C in sea water and 1°C-22°C in compost). Only the average temperature of sea water and compost during summer months (July, August) was on the similar level to the preferred for enzymatic degradation (20°C-60°C) [30].

There were significant differences in the pH values of both natural environments. The average pH in the Baltic Sea was 8.2 and 5.5 in compost pile.

The very low temperature and the alkaline pH (~8) of the Baltic Sea caused that the psychrotrophic bacteria could play the main role in the degradation in this environment [25].

During the winter months, we could observe the very low temperature and the highest oxygen content (February 10, 3 cm^3/dm^3). These conditions could have an influence on the activity of oxidizing enzymes, which are responsible for oxidation. It could be expected that these conditions had an influence on the development of aerobic epilithic bacteria. The metabolism of these microorganisms probably caused the decrease of oxygen content in the summer months (August 6.5 cm^3/dm^3) and changed the concentration of carbon dioxide in sea water. However, it was not able to change significantly the pH of sea water.

It is known that the activity of dehydrogenases depends on the degree growth of microorganism populations, which are producing enzymes involved in degradation process. During the degradation time, the activity of dehydrogenases had been changing and depending on both biotic and abiotic conditions in this environment.

The weather, as well as respiration of microorganisms, had an influence on fluctuation of moisture content in the compost. With decreasing moisture content, lower absolute value of the activity of dehydrogenases was observed.

The rather low temperatures (below 20°C) and slightly acid pH (~6) of compost under natural weather - depending conditions caused that psychrotrophic acidophilic microorganisms (fungi) could play the main role in the degradation process [26].

Considering the characteristic abiotic parameters of sea water and compost presented in Tables 1 and 2 and the different microbial communities (fungi in compost and bacteria in sea water), we could expect the different rate of degradation of polyurethanes in these two natural environments.

Incubation of polyurethane samples in the laboratory in a liquid medium containing sea water with NaN_3 (Table 2) was performed in a stable temperature (about 20°C) and under alkaline conditions (pH about 8).

3.2. Evaluation of the changes of polyurethane samples during environmental degradation

The environmental degradation of polyurethanes in sea water and compost was evaluated visually at first. Figure 4 represents the surface view at macro scale of investigated poly(ester-urethanes) before and after degradation in the Baltic Sea water and compost under natural weather depending conditions.

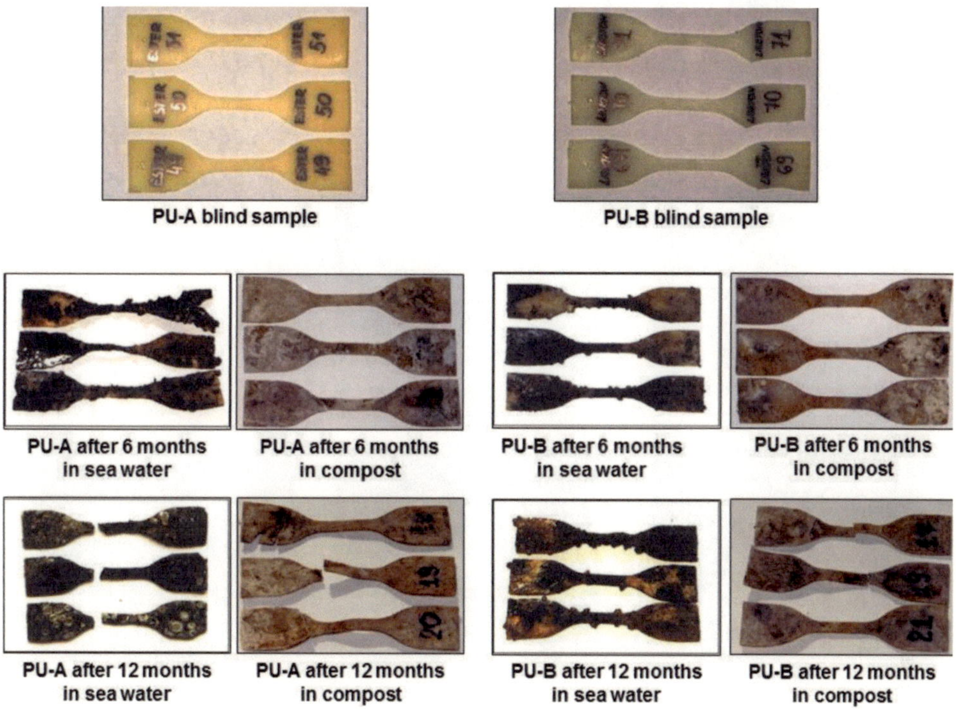

Figure 4. Macroscopic images of poly(ester-urethanes) after environmental degradation.

The blind samples of PU-A and PU-B are beige or white and opaque. The both poly(ester-urethane) samples incubated in natural environments are characterized by the flaws at the surface that gradually grew into microcracks, eventually breaking the samples. At the end of the experiment in natural environments, the surface of poly(ester-urethanes) samples is rough and cracked with visible black areas after incubation in sea water. While after incubation in sea water with NaN_3 the surface is only matt.

The environmental degradation of poly(ester-urethane) samples in the Baltic Sea water and the compost under natural weather-depending conditions was also evaluated on the basis of changes of surface morphology. After incubation in both natural environments, the poly(ester-urethane) samples were not homogeneously destroyed over the whole polymer surface and there were different images depending on the place. The photomicrographs repeated images observed under the reflected microscope equipped with polarizer were done.

Microscopic observations after incubation in the natural environments have shown vulnerability of PU-A and PU-B to the microbiological attack. The changes of surfaces of both poly(ester-urethanes) are comparable. After incubation of poly(ester-urethanes) in natural environments the deterioration of the surfaces has been observed. The photomicrographs of PU-B are presented in Figure 5.

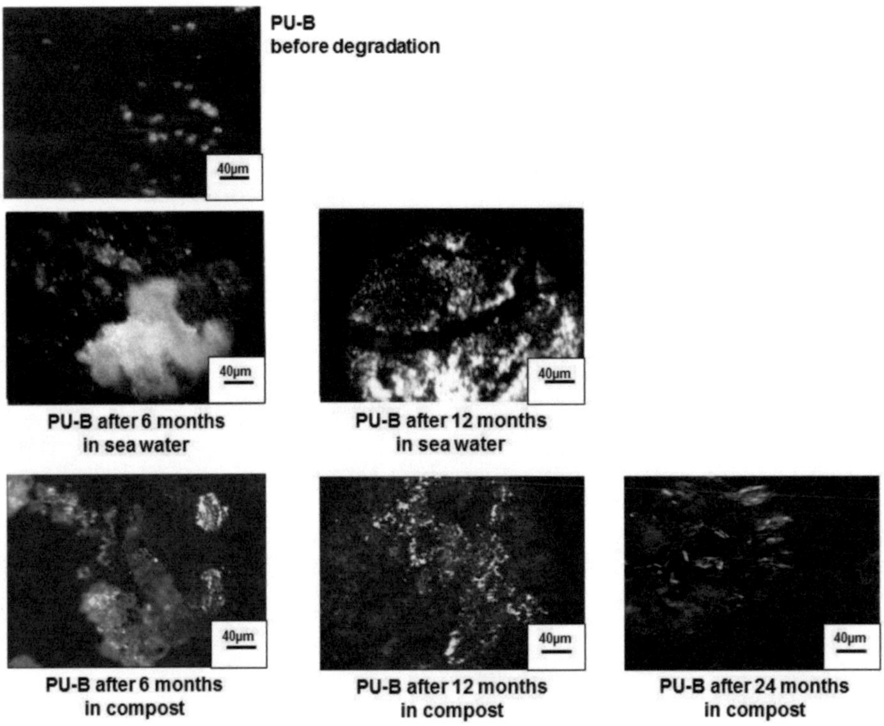

Figure 5. Microscopic images of PU-B observed under optical microscope with polarizer before and after environmental degradation.

The changes in birefringence of PU-B observed under a polarized optical microscope are noticeable (Figure 5). The blind sample of PU-B is slightly crystalline (the bright elements on the surface before degradation). It is known that the degradation process of polyesters proceeds in two stages [31]. During the first stage of the environmental degradation of PU-B, the gradually increase of bright elements on the surface was observed. This increase of bright elements might be an evidence of increase in crystallinity as a result of the degradation amorphous phase (random hydrolytic scission of ester bonds) [25, 26]. The second stage started after degradation of the major part of the amorphous phase. At the end of incubation in the compost, we could see distinct loss of bright elements due to the degradation of the crystalline phase (Figure 5).

Comparing microscopic observations of PU-B after environmental degradation in the Baltic Sea water and in compost we can state that PU-B based on poly(ε-caprolactone) is more vulnerable to degradation in compost than in sea water. Thus, the higher deterioration of the PU-B surface have been observed. It is because of the fragment of poly(ε-caprolactone) in the main chain of PU-B, which is susceptible to microbial degradation. Moreover, in this natural environment there were conditions favourable for growth of fungi. It is well known that fungal communities are involved in poly(ester-urethanes) biodegradation [2, 6, 17, 23, 24, 32].

Susceptibility of poly(ester-urethanes) to environmental degradation was evaluated based on weight changes [%] of polymer samples after incubation. The results of the weight changes of the PU-A and PU-B after incubation in natural environments are presented in Figure 6.

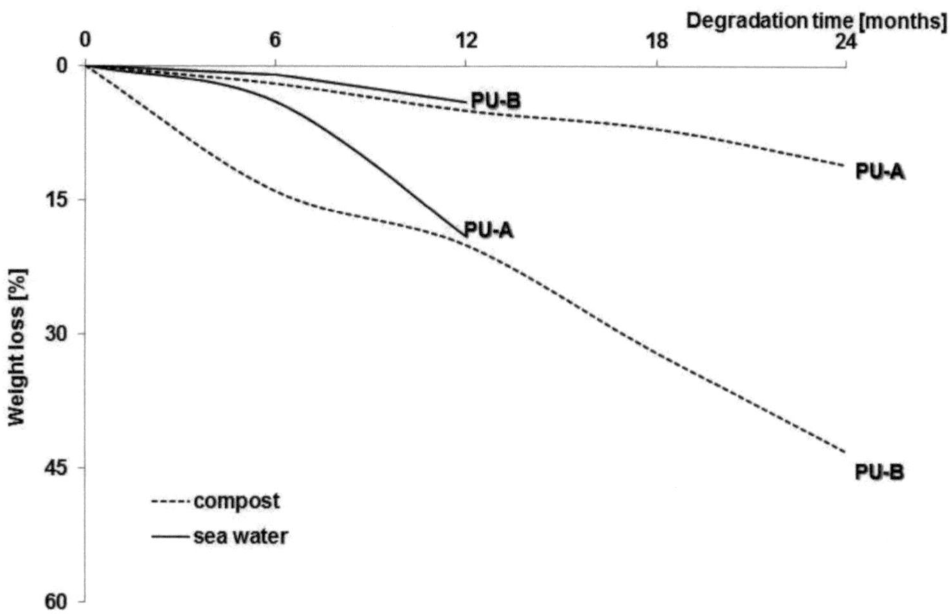

Figure 6. The weight losses [%] of PU-A and PU-B after environmental degradation.

The obtained results indicate that both poly(ester-urethanes) degrade in natural environments such as the Baltic Sea water and compost. It confirms susceptibility of poly(esters-urethane) to biological degradation.

Generally, the results presented in Figure 6 indicate that the degradability of poly(ester-urethanes) depends on their chemical structure and the kind of environment.

In the Baltic Sea water, where the low temperature and the alkaline pH were favourable for the development of aerobic bacteria, uncrosslinked PU-A based on poly(ethylene-butylene-adipate) was more prone to degradation than PU-B based on poly(ε-caprolactone). At the end of incubation in the Baltic Sea water (12 months) the changes of weight of PU-A were much higher (19%) than of PU-B (4%). It could be explained by the different networks of those polyurethanes. Unexpectedly, the introduction of a fragment of poly (ε -caprolactone) to the main chain of PU-B did not lead to the increase of its environmental degradability [25].

In contrary, to this there are results of the degradation of poly(ester-urethanes) in compost under natural weather-depending condition. The strongest effect of environmental degradation in compost was observed for slightly crosslinked PU-B based on poly (ε -caprolactone) than not crosslinked PU-A based on poly(ethylene-butylene-adipate). It could be mainly explained by degradation of poly (ε -caprolactone) in this biotic environment as a result of enzymatic hydrolysis of ester bonds susceptible to fungal degradation [26]. Considering the parameters of compost pile under natural weather-depending conditions (temperature and pH) it could be stated that the psychrotrophic acidofilic microorganisms (fungi) were responsible for the level of degradability of poly (ester-urethanes).

According to the literature polyester degradation occurs primarily by chain scission in main chain of polymer and can be induced by enzymatic hydrolysis. Enzymes can attack on the surface poly(ε-caprolactone) segments of polyurethane, degrading them to smaller molecular units via hydrolytic attack. Then surface erosion takes place with farther hydrolysis process erosion. At the end, hydrolysis rate decreases after the consumption of the amorphous materials by microorganisms [6, 33, 34].

This is why the higher weight losses and deterioration of the surface have been observed during environmental degradation in compost [26].

During incubation in the sea water with NaN_3 in the laboratory, the weight changes of poly(ester-urethane)s are insignificant even though the temperature was higher than in the natural environment. This was due to the absence of microorganisms in sea water with sodium azide [25].

Changes of mechanical properties of both poly(ester-urethanes) were checked by the measurement of the tensile strength before and after environmental degradation. The results are presented in Figure 7.

It is interesting to note that for blind samples, the higher tensile strength is observed for PU-B sample, which is due to its partial crosslinking. The rates of the changes in the mechanical properties (Figure 7) resemble the rate of the mass loss (Figure 6) and the changes of surface poly(ester-urethanes) (Figures 4 and 5). The data in Figure 7 show that the tensile strength of poly(ester-urethanes) had been decreased during the incubation time in sea water and compost [25, 26]. After 6 months of environmental degradation, only the samples of PU-A incubated in sea water were torn up into pieces, whereas the tensile strength of the other samples degraded in both environments could still be estimated. Probably, microorganisms existed in sea water, such as psychrotrophic bacteria, caused the breaking of polymer samples resulting in the fragmentation. After 12 months of environmental degradation all poly(ester-urethanes) lost the tensile strength.

The loss of tensile strength, discoloration, and cracking observed for environmental degraded poly(ester-urethanes) are typical for the effects of degradation of poly(ester-urethanes) as a result of microorganisms activity.

The results of thermal analysis of poly(ester-urethane) samples are shown in Table 4 and Figure 8a and b and they are in the agreement with microscopic observations (Figure 5).

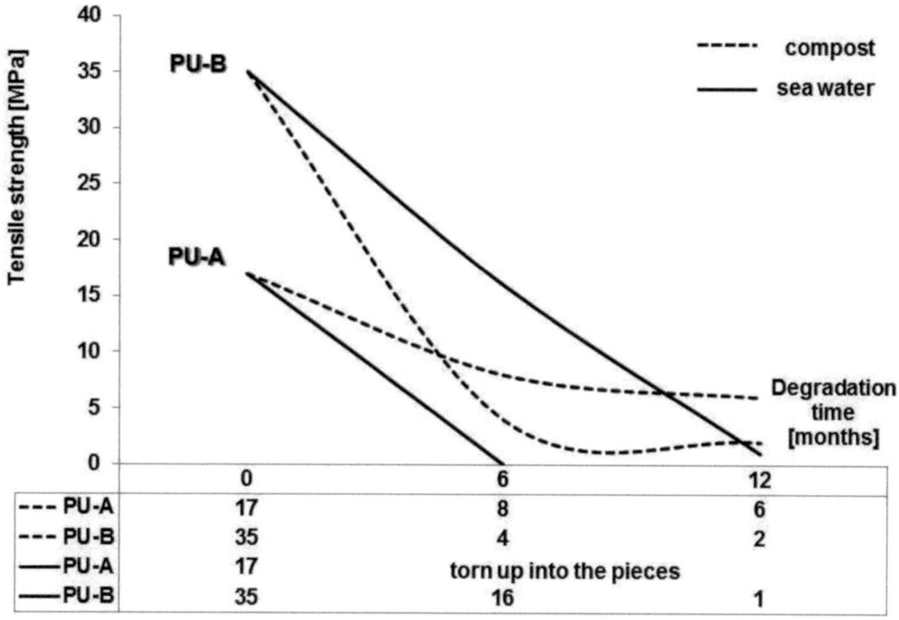

Figure 7. The tensile strength [MPa] of PU-A and PU-B before and after environmental degradation.

Polymer	Incubation time [months]	Tm_{SS} [°C]	ΔHSS [J/g]	Tm_{HS} [°C]	ΔH HS [J/g]
PU-A	0	-	-	140/184	3/3
	6	48	8	146	39
	12	51	9	136	2.3
	24	55	8	151	22
PU-B	0	77	2.6	-	-
	6	49	5.5	188	3
	12	50	3	143/196	2/1
	24	52	7.6	189	2

Table 4. Melting temperature (Tm) and melting enthalpy (ΔH) of crystallites made of soft (SS) and hard segment (HS) determined from DSC scans for PU-A and -B samples before and after incubation in compost.

The DSC analysis of PU-A and PU-B revealed the differences in their phase composition (Figure 8). Both poly(ester-urethanes) before degradation showed the little presence of crystal phases. The blind sample of PU-A contains mainly crystals made of hard segments, which is indicated by the small melting peaks at high temperatures (140°C and 184°C in Figure 8a). In the case of the blind sample of PU-B, only small melting peak at low temperatures (77°C in Figure 8b) was noticed. It is corresponding to the melting of soft segments crystals.

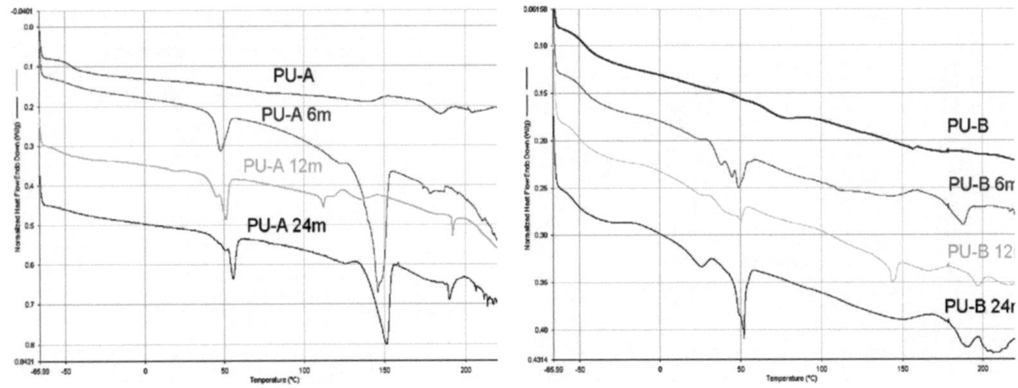

(a) (b)

Figure 8. The DSC curves for poly(ester-urethanes): **a)** PU-A and **b)** PU -B before and after incubation in the compost for 6, 12, and 24 months [26]

Due to incubation in the compost, the evident increase in crystallinity in both poly(ester-urethanes) was observed. However, in the case of PU-A, crystal phases made of both hard and soft segments appeared, but in the case of PU-B, mainly crystal phase made of soft segments (compare Figure 8a and b and Table 4) appeared. The observation that higher crystallinity develops in PU-A may be explained by its uncrosslinked structure. In the case of PU-B, the partial crosslinking constraints at some point of crystallization are especially of hard segments. Differences in crystallinity in both poly (ester-urethanes) seem to correspond with mechanical properties. The smaller decrease of mechanical properties of PU-A is due to the reinforcing effect of higher crystallinity [26].

4. Conclusions

Currently, the information concerning microbial degradation of polyurethanes in the natural environment is still limited. In this study, the ability of the Baltic Sea water and compost to degrade poly(ester-urethanes) were accessed. The achieved results pointed out that the poly(ester-urethane) based on poly(ethylene-butylene-adipate) and poly(ester-urethane) based on poly (ε -caprolactone) are susceptible to environmental degradation under natural weather-depending conditions.

Generally the stronger effect of environments degradation—the higher weight losses and deterioration of the surface of poly(ester-urethanes)—has been observed during environmental degradation in compost, than in sea water where the conditions were favourable for the development of aerobic bacteria.

In the Baltic Sea water uncrosslinked poly(ester-urethane) based on poly(ethylene-butylene-adipate) was very prone to degradation. Whereas slightly crosslinked poly(ester-urethane) was moderately resistant to degradation in sea water, even though it has a fragment of poly (ε -caprolactone) in the main chain. In contrary to this, there are results of the degradation of poly(ester-urethanes) in compost. In this environment, slightly crosslinked poly(ester-urethane)B, based on poly(ε-caprolactone) was more degradable, than not being crosslinked poly(ester-urethane)A, based on poly(ethylene-buthylene adipate). Finally, the higher weight losses and deterioration of surface have been observed. It could be mainly explained by degradation of poly (ε -caprolactone) in this biotic environment as a result of enzymatic hydrolysis of ester bonds susceptible to fungal degradation. The psychrotrophic acidofilic microorganisms (fungi) were responsible for the level of degradability of poly(ester-urethanes) in compost.

Due to incubation in the compost, there is the evident increase in crystallinity in both poly(ester-urethanes). Differences in crystallinity are corresponding with mechanical properties as reinforcing effect of crystal phase in poly(ester-urethanes). The smaller decrease in mechanical properties of poly(ester-urethane)A, than poly(ester-urethane)B may be due to its higher crystallinity. The loss of tensile strength, discoloration, and cracking observed for environmental degraded poly(ester-urethanes) are typical for the effects of degradation of poly(ester-urethanes) as a result of microorganisms activity.

The environmental degradation process of poly(ester-urethanes) indicates that the degradation in natural environment such as the Baltic Sea water and compost is the result primarily of enzymatic hydrolysis of ester bonds, then crystallinity and network structure of poly (ester-urethanes). The rate of environmental degradation process of poly(ester-urethanes) also is depended on the kind and conditions of natural environment.

Author details

Katarzyna Krasowska, Aleksandra Heimowska and Maria Rutkowska*

*Address all correspondence to: m.rutkowska@wpit.am.gdynia.pl

Gdynia Maritime University, Faculty of Entrepreneurship and Quality Science, Department of Chemistry and Industrial Commodity Science, Gdynia, Poland

References

[1] Wood G, editor. The ICI polyurethanes book. 2nd ed. New York: Wiley & Sons Inc.; 1990.

[2] Howard GT. Biodegradation of polyurethane: A review. International Biodeterioration & Biodegradation. 2002;49:245-252. DOI: 10.1016/S0964-8305(02)00051-3.

[3] Zhang Y, Xia Z, Huang H, Chen H. A degradation study of waterborne polyurethane based on TDI. Polymer Testing. 2009;28:264-269. DOI: 10.1016/j.polymertesting.2008.12.011.

[4] Umare SS, Chandure AS. Synthesis, characterization and biodegradation studies of poly(ester urethane)s. Chemical Engineering Journal. 2008;142:65-77. DOI: 10.1016/j.cej.2007.11.017.

[5] Gao X, Guo Y, Tian Y, Li S, Zhou S, Wang Z. Synthesis and characterization of polyurethane/zinc borate nanocomposites. Colloids Surfaces A-Physicochemical and Engineering Aspects. 2011;384:2-8. DOI: 10.1016/j.colsurfa.2010.11.037.

[6] Cometa S, Bartolozzi I, Chellini F, Giglio E, Chiellini E. Hydrolytic and microbial degradation of multi-block polyurethanes based on poly(caprolactone)/poly(ethylene glycol) segments. Polymer Degradation and Stability. 2010;95:2013-2021. DOI: 10.1016/j.polymdegradstab.2010.07.007.

[7] Pretsh T, Jakob I, Muller W. Hydrolytic degradation and functional stability of a segmented shape memory poly(ester urethane). Polymer Degradation and Stability. 2009;94:61-73. DOI: 10.1016/j.polymdegradstab.2008.10.012.

[8] Yeganeh H, Hojati-Talemi P. Preparation and properties of novel biodegradable polyurethane networks based on castor oil and poly(ethylene glycol). Polymer degradation and Stability. 2007;92:480-489. DOI: 10.1016/j.polymdegradstab.2006.10.011.

[9] Lelah MD, Cooper SL. Polyurethanes in medicine. Boca Raton, Florida: CRC Press Inc.; 1986.

[10] Planck H, Syre I, Dauner M, Egbers G, editors. Polyurethane in biomedical engineering. 2nd ed. Amsterdam: Elsevier Science Ltd; 1987.

[11] Cherng JY, Hou TY, Talsma H, Hennik WE. Polyurethane-based drug delivery system. International Journal Of Pharmaceutis. 2013;450:145-162. DOI: 10.1016/j.ijpharm.2013.04.063.

[12] Ding M, Li J, Tan H, Fu Q. Self-assembly of biodegradable polyurethanes for controlled delivery applications. Soft Matter. 2012;8:5414-5428. DOI:10.1039/C2SM07402H.

[13] Guelcher SA, Srinivasan A, Dumas JE, Didier JE, McBride S, Hollinger JO. Synthesis, mechanical properties, biocompatibility and biodegradation of polyurethane networks from lysine polyisocyanates. Biomaterials. 2008;29:1762-1775. DOI: 10.1016/j.biomaterials.2007.12.046.

[14] He X, Zhai Z, Wang Y, Wu G, Zheng Z, Wang Q, Liu Y. New method for coupling collagen on biodegradable polyurethane for biomedical application. Journal of Applied Polymer Science. 2012;126:E353-E360. DOI: 10.1002/app.36742.

[15] Zdrahala RJ, Zdrahala IJ. Biomedical applications of polyurethanes: A review of past promises, present realities, and a vibrant future. Journal of Biomaterials Applications. 1999;14:67-90. DOI: 10.1177/088532829901400104.

[16] Young RJ, Lovell PA, editors. Introduction to polymers. 2nd ed. London: Chapman & Hall; 1994.

[17] Amaral JS, Sepulveda M, Cateto CA, Fernandes IP, Rodrigues AE, Belgacem MN, Barreiro MF. Fungal degradation of lignin-based rigid polyurethanes foams. Polymer Degradation and Stability. 2012;97:2069-2076. DOI: 10.1016/j.polymdegradstab.2012.03.037.

[18] Shah AA, Hasan F, Hameed A, Ahmed S. Biological degradation of plastics: A comprehensive review. Biotechnology Advances. 200228;26:246-265. DOI: 10.1016/j.biotechadv.2007.12.005.

[19] Eubeler JP, Bernhard M, Knepper TP. Environmental biodegradation of synthetic polymers II. Biodegradation of different polymer group. Trends in Analytical Chemistry. 2010;29:84-100. DOI: 10.1016/j.trac.2009.09.005.

[20] Gorna K, Gogolewski S. In vitro degradation of novel medical biodegradable aliphatic polyurethanes based on caprolactone and Pluronics with various hydrophilicities. Polymer Degradation and Stability. 2002;75:113-122. DOI:10.1016/S0141-3910(01)00210-5.

[21] Khan I, Smith N, Jones E, Finch DS, Cameron RE. Analysis and evaluation of a biomedical polycarbonate urethane tested in an in vitro study and an ovine arthroplasty model. Part I: Materials selection and evaluation. Biomaterials. 2005;26:621-631. DOI: 10.1016/j.biomaterials.2004.02.065.

[22] Urgun-Demirtas M, Singh D, Pagilla K. Laboratory investigation of biodegradability of a polyurethane foam under anaerobic conditions. Polymer Degradation and Stability. 2007;92:1599-1610. DOI: 10.1016/j.polymdegradstab.2007.04.013.

[23] Lu H, Sun P, Zheng Z, Yao X, Wang X, Chang FC. Reduction-sensitive rapid degradable poly (urethane-urea)s based on cystine. Polymer Degradation and Stability. 2012;97:661-669. DOI: 10.1016/j.polymdgradstab.2011.12.023.

[24] Tatai L, Moore TG, Adhikari R, Malherbe F, Jayasekara R, Griffiths I, Gunatillake PA. Thermoplastic biodegradable polyurethanes: The effect of chain extender structure on properties and in vitro degradation. Biomaterials. 2007;28:5407-5414. DOI: 10.1016/j.biomaterials.2007.08.035.

[25] Rutkowska M, Krasowska K, Heimowska A, Steinka I, Janik H. Degradation of polyurethanes in sea water. Polymer Degradation and Stability. 2002;76:233-239. DOI: 10.1016/S0141-3910(02)00019-8.

[26] Krasowska K, Janik H, Gradys A, Rutkowska M. Degradation of polyurethanes in compost under natural conditions. Journal of Applied Polymer Science. 2012;125:4252-4260. DOI: 10.1002/app.36597.

[27] Rutkowska M, Balas A. Influence of NCO:OH relation and chain extender on primary and secondary valence crosslinks in urethane elastomers. Journal of Applied Polymer Science. 1980;25:2531-2538. DOI: 10.1002/app.1980.070251108.

[28] Rutkowska M, Eisenberg A. Ionomeric blends. IV. Miscibility of urethane elastomers with styrene-styrene sulfonic acid copolymer. Journal of Applied Polymer Science. 1984;29:755-762. DOI: 10.1002/app.1980.070290304.

[29] PN-EN ISO 527-1,2,3:1998, Plastics—Determination of Tensile Properties.

[30] Lenz RW. Biodegradable Polymers. In: Advances in Polymer Science. Berlin: Springer-Velay; 1993. p. 1-40.

[31] Li S, Vert M. Biodegradation of Aliphatic Polyesters. In Degradable Polymers Principles and Applications. 2nd ed. UK: Kluwer Academic Publisher; 2002. p. 71-132. DOI: 10.1007/978-94-017-1217-0.

[32] Barrat SR, Ennos AR, Greenhalgh m, Robson GD, Handley PS. Fungi are predominant micro-organisms responsible for degradation of soil-buried polyester polyurethane over a range of soil water holding capacities. Journal of Applied Microbiology. 2003;95:78-85. DOI: 10.1046/j.1365-2672.2003.01961.x.

[33] Zhou L, Liang D, He X, Li J, Tan H, Fu Q, Gu Q. The degradation and biocompatibility of pH-sensitive biodegradable polyurethanes for intracellular muntifunctional antitumor drug delivery. Biomaterials. 2012;33:2734-2745. DOI: 10.1016/j.biomaterials.2011.11.009.

[34] Khan S. Biodegradation of polyurethane by bacterial consortium. Elixir Bio Technology. 2011;37:3767-3772.

Chapter 5

Heat sensing Thermoplastic Elastomer Based on Polyolefins for Encapsulation Applications

Tanya Das and Sunanda Roy

Additional information is available at the end of the chapter

http://dx.doi.org/10.5772/61691

Abstract

Use of Thermoplastic Elastomers (TPEs) has become a unique pathway to meet the daily requirements of various applications. The ease of using TPEs lies in the fact that they provide both the character of the individual properties as they are constructional polymers, which are physically crosslinked materials made up of a thermoplastic and an elastomer. There are several TPE's in market and individual have several outstanding performances. Out of several researches, our aim in this article is to focus on the influence of Polyolefin based TPE's. This paper focusses on the different aspects of TPO's their physical, chemical, mechanical, and electrical characteristics, advantages and uses of these materials along with a particular focus on their use in encapsulation application. Factors that could affect the end use are also explained here in details. Heat shrinkability test, cure time, and SEM are some of the characterisations used to demonstrate the exact criteria of polyolefin based TPE's for encapsulation application.

Keywords: Thermoplastic elastomer (TPE), Thermoplastic polyolefins (TPO's), Shrinkability, Encapsulation application

1. Introduction

Elastomers are polymers which have viscoelasticity and weak intra-molecular forces. They can be generally explained by their mechanical response rather than their chemical structure. One of the most versatile and immensely used materials in today's world is the engineered Thermoplastic elastomers (TPE's). The TPE's are a class of polymer which consists materials having both thermoplastic and elastomeric properties. They are novel constructional polymers, which are physically cross linked materials made up of a thermoplastic and an elastomer. The unique properties of both materials exist because TPE materials are created only by physical mixing of a thermoplastic and an elastomer and no chemical or covalent bonding exists between the two [1]. TPEs are generally low modulus, flexible materials that can be stretched repeatedly to at least twice their original length at room temperature with an ability to return to their approximate original length when stress is released. TPE's offer a wide range of performance attributes including: heat and oil resistance, improved adhesion, tear resistance, surface appearance and low permeability. In addition, TPE's are colorable and can be specified in a variety of hardness grades. They can be processed with the efficiency and economy of thermoplastics and can be molded with other olefin based materials, such as polypropylene, polyethylene without the use of adhesives. Various mechanical designs were implanted within the parts to ensure tight fit in case of other substrates like polyamides, acrylonitrile butadiene styrene and so on. Rubbery materials have high degree of flexibility in their elongated polymeric chains which binds them into an ordered network structure [1]. Due to this mobility and flexibility the long chains when subjected to external pressure or stress change their configuration. Thus chain allied to a structure gives a solid feature preventing them to flow under pressure. Due this fact, this type of materials may be stretched or pulled up to several times of its original length. Withdrawing external forces, it rapidly restore its nearly original dimensions, with essentially no residual or nonrecoverable strain. TPEs serve a wide range of markets:

- Agriculture & Off Road Appliance
- Automotive & Transportation · Consumer
- Electrical & Industrial Controls · Food & Beverage
- Hydraulics & Pneumatics · Marine
- Medical & Safety Plumbing & Irrigation

One of the major usages of TPE is in encapsulations. By encapsulation we mean covering of an object with some material to protect and to insulate it. Polymeric materials are used extensively as encapsulants for coils, motor windings and micro electronics packaging industry. Use of TPEs speed manufacturing with low cost compared with any other material and technology. TPE gives both the characteristics like thermosets along with the process and design flexibility same as plastics thereby enhancing wide design alternatives and cost reduction facility. TPE's can be processed faster more easily and more efficiently. Apart from these advantages TPE's possess many other additional characteristics like lightweight, elastic recovery properties within a specific temperature range, metal replacement, noise reduction by self lubrication, very good electrical insulation properties, heat resistance with in a specific temperature range, oil resistance, improved adhesion, tear resistant surface, low permeability and colorability [2]. TPE's which have contribution in encapsulation application are Polystyrene, Polyolefin and Polyurethanes. In this chapter we will concern our focus on Polyolefins.

1.1. Compounding of TPE's

The basic characteristics of thermoplastic elastomer greatly depend on polymers used in their manufacturing process based on their application. These properties can be modified, however, through the appropriate addition of compounding ingredients. Some are added to accelerate

http://dx.doi.org/10.5772/61691

cross-linking in order to impart the exact cure time required some improve processability, while others improve the properties of the finished product.

Some ingredients are added to deliver the highest levels of performance in the end product. In some applications, compounding is required to reduce cost, diluents and extenders are sometimes added to minimize the ratio of high valued components within the system. By this, certain compromises in the mechanical and other properties are obtained, but for some specific high end applications it is worthwhile. Some other factors that influence the quality of the resulting material includes the quality of the neat materials, the equipment's used and the quality control in mixing techniques [3]. Processing of the compounded material can also influence the end product properties. The major ingredient that affect the properties of the finished product and which have its own importance while making an end product are classified as:

1.1.1. Matrix

The neat or raw polymer, is the fundamental or main ingredient in determining the properties of the compound or the end product. It is selected in such a way so as to optimize service life and processing requirements with cost effectiveness as one of the main parameter taken into account. Polymers with high molecular weight can give very tough material as the final product. Again this criterion sometimes becomes a disadvantage for some specific product. So depending on the end use application selection of polymer is yet challenging.

1.2. Filler

The most important part of filler is to provide better reinforcement and secondly to reduce the cost of the end product. Generally we have two basic types of fillers. One is the reinforcing type which can reinforce the system according to need and the other is the diluent type which are generally used for calibrating the physical properties of the system. Carbon Black or Nano silica's basically come under reinforcing fillers and are categorized by their respective particle sizes. They become more reinforcing as the particle size decreases [4]. As mentioned above fillers with high reinforcing character can make a compound very hard and rigid which in turn can result in poor flow behavior. Carbon blacks are generally alkaline in nature and tend to accelerate cure time. Other non-black fillers may be acidic and can retard cure as well as absorb moisture, which can result in blistering problems during the processing stage. Several nano-fillers like super-fine clays have a high surface area as compared to their volume and can induce better mechanical behavior. They are more expensive than general fillers, the same weight of material goes further because the particles are so much finer.

1.2.1. Accelerators

These speed up the cure mechanism of the system. Accelerators increase the rate of cross-linking and therefore decrease both production time and cost.Accelerators can help to control the cure speed and cure time and also other material properties. They can be incorporated within the system at any amount and sometimes may be more than one accelerators are used in a single formulation. Some important accelerators include peroxides for curing which plays dual character in the network. They can act like an accelerator as well as a modifier for the physical properties of the system.

1.2.2. Flame retardants

These are compounds added to materials to suppress or delay the flames and prevent fire spreading. Most of the TPEs promote combustion and thus the end by-products can be extremely hazardous and dangerous. Due to this fact manufacturers induce flame retardants to improve their flame resistance. There are several flame retardants that can be added to the compound, either inorganic or organic. These include antimony trioxide, zinc borate, aluminum hydroxide and chlorinated paraffins and so on.

1.2.3. Curatives

Curatives are very crucial for compounding of elastomers. They are the crosslinking agents generally used to connect separate neat polymers. The type of curative used varies upon the elastomer selected. Like in sulfur-cured, sulfur donors give better heat stability as they tend to give single sulfur crosslinks. Peroxide cures promotes good thermal stability due to the short length of the cross-links between the polymer chains. Fluorocarbons, along with some other polymer types, can have their own specialized cure systems.

1.2.4. Plasticizers

Plasticizers help to reduce the hardness with given filler loading and also gives better filler addition and dispersion. They should be compatible with the matrix polymer. Special types of plasticizers can improve the low temperature flexibility of some rubber. Process aids can also assist with filler dispersion, although they are normally added to improve processability.

2. General characteristics of elastomers

2.1. Cure characteristics

The cure characteristic is one of the major properties of an elastomer to study before its application to different areas. As the compound cures within the hot platens it gradually becomes stiffer. This is measured via a strain gauge connected to an oscillating rotor in contact with the elastomer. The resistance to torque or the stiffness of the material is plotted on a graph against time, known as a rheograph. This information shows the behavior of the moulding characteristics, since the rheograph shows the time available to load the press, the time of cure and the final state of cure. The ultimate state of cure is not always a straight line but can also be a slope [5]. For some elastomers the cure continues in a different way which could be explained as the heat is actually breaking the polymer chain rather than the crosslinks formed during the curing stage.

http://dx.doi.org/10.5772/61691

2.2. Abrasion resistance

Abrasion resistance is important where friction, rough surface, industrial use is a key factor. Hard TPOs have good abrasion resistance. Abrasion damage can occur when there is dynamic motion against an abrasive surface. Standard abrasion tests depend on producing relative motion between a rubber sample and an abrasive surface, pressed together by a fixed force. These tests do not correlate particularly well with application experience. It is sometimes believed that tensile strength is related to abrasion resistance, higher the tensile strength higher is the abrasion resistant property while a high tensile strength compound can have good abrasion resistance. Abrasion resistance is mainly related more to the polymer taken and the nature of compounding ingredient used [6]. Abrasion resistant elastomers must therefore be specifically developed. Polyolefin based TPEs show good abrasion resistant character.

2.3. Tear strength

Tear strength is a measure of the resistance of an elastomer to tearing. It is measured using a tensile test machine operating at a constant rate of traverse until the test piece breaks. Various types of test pieces can be used, and depending on the method employed the maximum or median force achieved is used to calculate the tear strength.

2.4. Adhesion and bondability

Parameters affecting the adhesion and bond ability: The application of elastomers at high temperature is generally determined by their chemical structure and stability and varies depending on the elastomers being used. In course of developing high temperature product, these elastomers can be attacked by several chemical species like hydrogen, oxygen, and other groups which results in some chemical reaction and as a result the effectiveness or power increases at high temperature. Another parameter responsible for affecting the adhesion and bond ability is the degradative chemical reactions. It may hamper the product in two specific ways. 1. Breaking the molecular chains or cross-links, thus softening the rubber because they weaken the network. 2. Additional cross-linking, hardening the rubber, often characterized by a hard, cracked and degraded skin occuring on the elastomer component [7]. One more important criterion responsible for adhesion and bond ability is the perfect selection of the materials based on the end product needed. For example elastomers significantly weaken at high temperatures so in the case of seals, can result in a significant reduction in extrusion resistance. For applications involving elevated temperatures, especially at high pressures, anti-extrusion elements may also need to be used, either incorporated into the seal design or added as an additional component when manufacturing the seal.

2.5. Low temperature applications

Elastomers when cooled to sufficiently low temperatures show the characteristics of glass, including hardness, stiffness and brittleness, and do not behave in the readily deformable manner usually associated with elastomers. As the temperatures rises, the segments of the polymer chain gain sufficient energy to rotate and vibrate. At high temperatures full segmental rotation is possible and the material behaves in the rubbery way. The usefulness of an elastomer at low temperatures is dependent on whether the material is above its glass transition temperature (Tg), where it will still behave elastically, or below its Tg, where the material will be hard and relatively brittle.

3. Polyolefins (POEs)

Polyolefins also named as poly alkenes are simple long chain hydrocarbon. They are prepared by the reaction of an alkene as a monmer with general formula C_nH_{2n}. The most commonly used Polyolefins are low and high density polyethylene copolymer, polypropylene copolymer and polymethyl pentene. Polyethylene and polypropylene are produced by long chain polymerization of olefins ethylene and propylene respectively

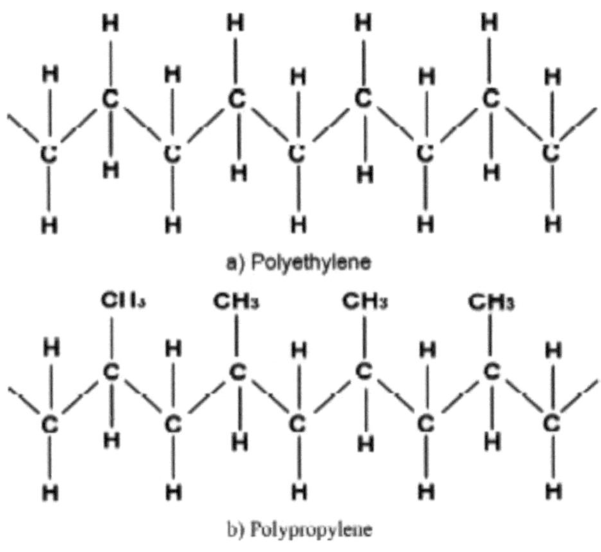

Sheme 1. Structure of Polyolefins

3.1. Chemistry and manufacturing process

The metallocene catalyst helps to instigate polymerization of the ethylene and co-monomer sequences and while further increasing the co-monomer content will produce polymers with higher elasticity due to the fact that the co-monomer incorporation disrupts the polyethylene crystallinity. The molecular weight of the copolymer will help determine its processing characteristics and end-use performance properties. Higher in molecular weights higher is the polymer toughness. Hence depending on the end use application one can choose the appropriate copolymer with molecular weight according to the requirement. POEs are produced using refined metallocene catalyst. These catalysts have a constrained transition metal such as

http://dx.doi.org/10.5772/61691

Ti, Zr, sandwiched between one or more cyclopentadienyl ring structures to form a sterically hindered polymerization site. The catalyst is usually first mixed with an activator or co-catalyst, which can significantly enhance the polymerization efficiency rate to beyond a million units of polymer per unit of catalyst [8-9]. The molecular weight build-up with the polymerization of ethylene and comonomer at the catalyst site until restricted by catalyst deactivation or chain termination with hydrogen introduction to the reactor. The polymerization process is exothermic in nature and requires efficient heat removal from the transport media of gas or solvent. Post-reactor processes involve additives addition and isolation of the polymer from the transporting media. The final product is then packaged as per manufacturer demand or end-user need.

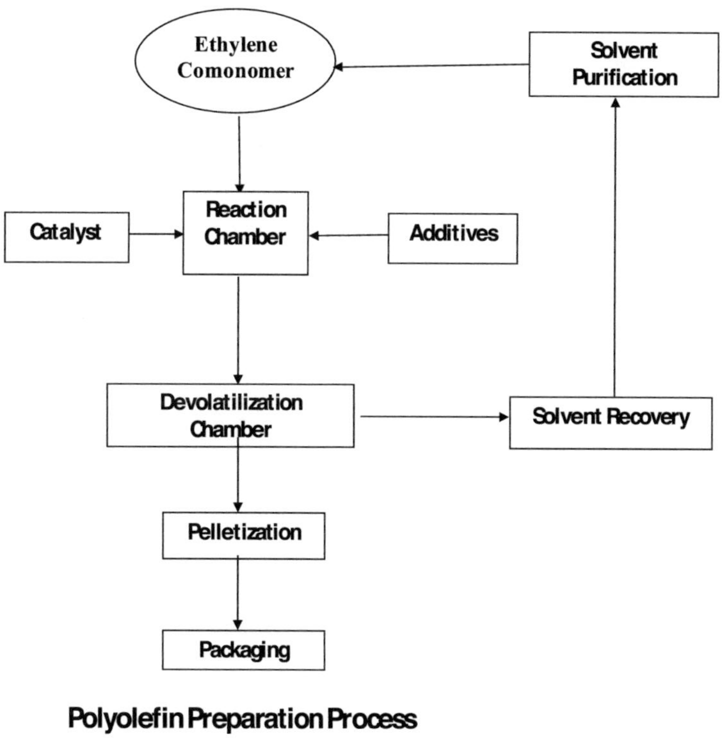

Sheme 2. Polyolefin Preparation Process

4. Thermoplastic Polyolefins (TPOs) blends

These are a class of plastic used in a wide range of markets and applications – mainly in transportation sector, including automotive exterior and interior parts. TPOs are generally produced by the blending of polypropylene (PP) with elastic ethylene copolymers (polyolefin elastomers or POEs), and the addition of other fillers and additives. Thermoplastic olefin (TPO) elastomers are obtainable in several grades in the market, having room-temperature hardness starting from 60 Shore A to 60 Shore D. These polyolefin elastomers have the lowest specific gravities of all thermoplastic elastomers available in the market. They are uncured or have low levels of crosslinking. These elastomers are flexible at around -60°F and are not brittle at -90°F. They can be used at service life at temperatures as high as 275°F in air. The TPOs have good resistance to some acids, most bases, butyl alcohol, many organic materials, formaldehyde, ethyl acetate and nitrobenzene. They are affected by chlorinated hydrocarbon solvents [10]. The specific blending amounts are dependent upon the properties needed to meet the desired end product application. TPO system generally include:

- Polymer., (provides rigidity and temperature stability to the body)
- Elastomers, which gives movability and impact strength to the system
- Filler (like minerals or talc), imparts higher stiffness and dimensional stability of the body
- Other additives (plasticizers, antioxidants, and additives for ignition resistance, scratch resistance) for improving end-use performance and shelf life.

Rigid TPOs are normally made up of polypropylene or polyethylene component in majority, with other component to attain an overall balance of properties. Efficient rigid TPO development starts by optimizing an appropriate polymer, and adjusting the optimum modifier level to achieve acceptable ductility, while maintaining rigidity as high as possible. In contrast, flexible TPOs contain a majority phase of elastomer with polymer added for improved temperature stability.

TPO's as one of the fastest growing synthetic polymers can be substituted for a number of generic polymers including ethylene styrene-block copolymers (SBCs), propylene rubbers (EPR or EPDM), ethylene vinyl acetate (EVA), and poly vinyl chloride (PVC). TPO's are well-matched with most of the olefinic materials, are an excellent impact modifier for plastics, and offer unique performance application for compounded products [11]. These compounds are easily colored, possess soft-touch molding and extrusion applications. They are generally resistant to acids and bases and are available in UV-resistant grades for outdoor applications. The main advantages are excellent abrasion resistance, high heat stability, wide use temperature, low-temperature flexibility and so on. Recently, TPO's has established itself as a leading material for automotive exterior and interior applications like extruded and molded goods, wire and cable, film applications, medical goods, adhesives, footwear, and foams. The advantage of using Polyolefins with other materials is that they have wide hardness range, excellent flex fatigue and impact resistance, good retention of properties etc. They are solvent resistant. The typical glass transition temperature can exceed 140C. They are also used to make lenses for cameras, projectors and copiers, LCD monitors, contact lenses for eye and reinforced tubes or pipes for medical use. TPO's provide an economical alternative to traditional thermoset rubbers and more costly thermoplastic elastomers [12]. Thermoplastic polyolefin elastomers provide an excellent balance of performance and price. Their service temperature range is from -60 to 275°F (-50 to 135° C); hardness ranges from 35 Shore A to 50 Shore D.TPO ingredients generally have a polymer (homopolymer, copolymer) which provides the rigidity and thermal stability, along with elastomers which gives the flexibility. It also includes mineral fillers which gives dimensional stability to the system. Other additive includes additives (antioxidants, plasticizers, and so on) depending on the application of the end use.

Sheme 3. Car Bumper made out of TPO material, (Source Google)

4.1. Heat sensing property of TP0's based materials

When a polymeric materials is subjected to stretch either at room temperature or at elevated temperature, the molecules of the materials are oriented towards the stretching direction and the material freezes in its elongated or extended form. Stretching decreases the entropy of the system. If we heat the stretched sample without applying mechanical force the material shrinks. Such heat shrinkable polymers are recently getting tremendous application in the packaging industry, cable industry and heat shrinkable tube production. As a flexible material elastomers do not provide the same level of tolerance as rigid materials do. Shrinkage is highly dependent on tolerance which again varies with the type of elastomers and other factors like hardness, stiffness and so on. Soft elastomers generally shrink more than harder elastomers. Shrinkage is also affected by other parameters like cure time, pressure, temperature, post-cure time, etc[13]. Most materials like metals, plastics or fabric have their own percentage tolerances. However, when designing inserts for moulding to elastomers, other factors need to be considered such as, the location of the inserts with respect to other shapes, fit in the mould cavities, proper spacing to match with mould pins, and the fact that inserts at room temperature should fit into a heated mould.

4.2. Electrical property of TPO's

With increasing demand for insulating materials researchers have paid attention towards materials that can provide the best according to the requirement. Various factors depend on to which materials can suit the best. Among them the most important criteria of a material as electrical insulation is determined by its dielectric breakdown strength. This in turn depends on some other properties like voltage applied, frequency, temperature, partial discharges, impurities in that material, dielectric constant, thickness and layers of insulation, and volume resistivity[14-15]. A material should fulfill some of the criteria to become useful in electrical insulation application. These properties include low specific weight, good mechanical, chemical and thermal strength, and good surface leakage resistance, ability to maintain good surface hydrophobicity, easy processing and production [4]. TPO gives a package of these combinations of properties and thus makes it most suitable to use in encapsulation application.. These materials are recently used widely for medium to high voltage cable encapsulations. They are also used in encapsulation of coils.

5. Experimental

5.1. Heat shrinkability test

a. Shrinkage was measured at 150°C for above materials.

b. Above materials were given stretching at ambient and at 150°C and then the shrinkage of the stretched samples was measured at the same temperature.

c. Above materials were stretched at 150C and cured under stretching conditions and then shrinkage was measured also at 150°C.

For R-T stretched samples the samples were stretched at room temperature at a stretching rate of 50 mm/min. Samples were allowed to shrink at 80 °C for 10 min in a hot air oven. The percentage shrinkability was measured by the following equation..

$$S_h = \left(L_{str} - L_{shr}\right)/L_{str} \times 100$$

where S_h is the percentage of shrinkability, Lstr is the length of the sample after stretching, and Lshr is the length of the shrunk sample.

Uncoiling of the polymer chains occurs when a polymer sample is stretched above the glass transition temperature, which in turn is frozen into the structure when the samples are cooled down. Although sufficient drop of entropy is witnessed during the uncoiling process the obtained stretched polymers are stabilized by a cohesive force received from the semicrystalline region of the polymers. Stretching also has some properties of its own. It causes unrecoverable viscous flow, which gradually increases with the increased stretching temperature and are not recoverable in any circumstances. Crosslinked points also known as permanent point strongly contribute in the shrinkage process. The molecules in the flexible amorphous phase shrink easily due to their mobility as compared to those in the crystalline zone. The observed shrinkability is a composite effect of the above mentioned factors. All samples were stretched at room temperature. The viscous flow is same in all cases.

5.2. Effect of cure time on heat sensing property of TPO

Some experiments were performed with different types of TPO's to study the heat sensing property. The parameter on which the heat sensing character depends gives a vast knowledge to the industries as well as other fields which deals especially with this parameter of TPO's.

http://dx.doi.org/10.5772/61691

Polymers taken are LLDE, LDPE, HPDE and the curing agents are mostly CSM, DCP, sulfur etc. The formulations are given in Table 1 and 2.

i	ii	iii
LLDPE/CSM/Sb_2O_3	LDPE/CSM/ Sb_2O_3	HDPE/CSM/ Sb_2O_3
40/60/1	40/60/1	40/60/1

Table 1. Formulation of TPO with curing agents

i	ii	iii	iv	v	vi
HDPE/BB/Sulfur	LDPE/BB/Sulfur	LLDPE/BB/Sulfur	HDPE/BB/DCP	LDPE/BB/DCP	LLDPE/BB/DCP
40/60/1	40/60/1	40/60/1	40/60/1	40/60/1	40/60/1

Table 2. Formulation of TPO with curing agents

Figure 1. elaborates that with increasing the cure time the shrinkability of the blends increases for all the systems. From the graph it could be observed that the LLDPE/CSM blends show the highest shrinkability degree both for Room Temperature and High Temperature stretched sample as compared to LDPE/CSM and HDPE CSM blends. H-T stretching is accompanied by high shrinkage rather than R-T stretched samples. Increase in cure time accelerates the shrinkability. This may be due to more crosslinking of the elastometric phase. In case of Figure 2. we notice the same trend as above. Shrinkability of all blends increases with increase in cure time. Here two types of curing agent were applied a) DCM b) sulfur. From both the figures it is apparent that rate of increase in shrinkability is much higher in case of DCP than sulfur as curing agent.

LLDPE/BB blends show the highest degree of shrinkability with increase with cure time for both the systems either with sulfur or DCP as the curing agent. But the extent of increase in shrinkability is even sharper in case of DCP as curing agent. This may be due to the fact that DCP attacks both polymer chains to generate active sites leading to more efficient crosslinking which results in higher degree of shrinkability whereas the sulfur only gives a bridging agent between the elastomers. Higher shrinkability of H-T stretched samples than the R-T stretched samples may be due to more crosslinking of elastomeric phase at elevated temperature to deposit curatives to crosslink further strongly which is not possible in case of room temperature stretched samples. Increase in cure time allows the elastomer phase to crosslink for longer duration resulting in sufficient crosslinking thereby increasing the shrinkability of the system.

5.3. Effect of flame retardant on shrinkability

Flame retardant pays an important role in shrinkability. It can be observed that with incorporation of some amount of flame retardants the shrinkability of the system is changed to some extent. For blends having high elastomer ratio the reduction in shrinkage occurs to a very small

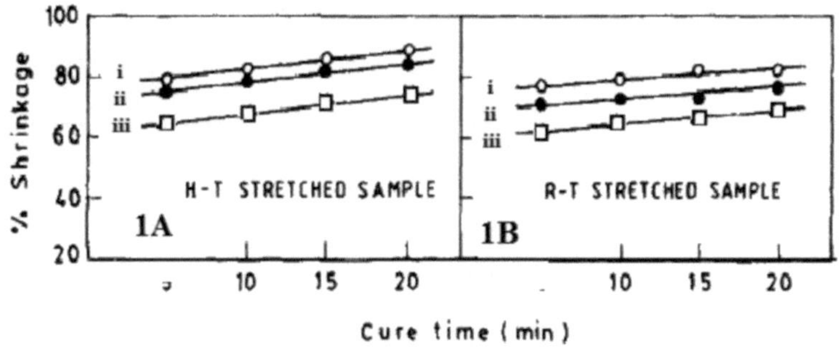

Figure 1. Study of shrinkability over cure time i) LLDBE/CSM, ii) LDPE/CSM, iii) HDPE/CSM

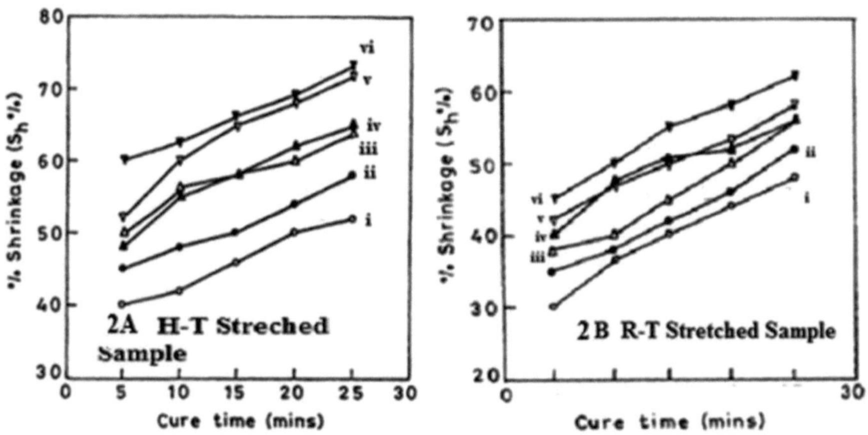

Figure 2. Study of shrinkability over cure time 2A) H-T stretched samples and 2B) R-T stretched samples of the blends of HDPE, LDPE, LLDBE with sulfur and DCP as curing agents.

extent in case of H-T stretched samples but for the low elastomer ration the shrinkability value is scarified to larger extent. The stretch ability of the H-T stretched blends again decreases to some extent with the addition of flame retardants. This may be the result of lower compatibility of flame retardants in the polyolefin phase at its higher level, suggesting the efficiency of flame retardants. The observed flame retardancy may be due to the fact that formation of some groups; entrap the flame propagating radicals like H, OH, O etc. and does not allow the flame to propagate. One more reason may be due to the fact that polymer structure after addition of flame retardants, is modified in such a way that char formation is facilitated thereby resisting the flame.

6. Encapsulation applications

Thermoplastic elastomeric encapsulation is used in many applications that require some special manufacturing techniques and materials of encapsulation. These include solenoids, sensors, self-supporting coils, transformers, motors, and electronic components of various types.

Various heat recoverable materials, which possess sufficient rigidity to hold out the elastomeric layer, may be used for the heat recoverable layer. Thermoplastic polymers are generally suitable heat recoverable materials either being crosslinked, or possessing the property of heat recoverability. Examples of such desirable TPEs that are crosslinked or which inherently possess the property of heat recoverability are polyolefins, such as ethylene-vinyl acetate copolymer, polyethylene, ethylene-ethyl acrylate copolymer or other ethylene copolymers, polyvinyl chloride, polyvinylidine difluoride, etc. Apart from that, other flexible polymers possessing necessary crystallinity such as trans-polybutadiene, ethylene-propylene-diene terpolymers, and trans-polyisoprene may also be used [16-18]. In addition, various commercially available elastomer-thermoplastic blends such as nitrile rubber-PVC and nitrile rubber-ABS may be taken into consideration. The only necessary property of the said thermoplastic material, used in place of the heat-recoverable material, should have sufficient rigidity at storage temperatures to seize the elastomeric sleeve. Such a material could be bonded to the elastomeric sleeve on either sides, by molding it in place or by placing concentric sleeves of thermoplastic material and relaxed elastomeric material in contact with each other and bonding them together at the interface.

In case of the elastomeric layer, any chosen material possessing elastomeric properties may be used. Suitable elastomers include rubber or rubber-like material such as natural rubber, styrene-butadiene rubber (SBR), cis-polyisoprene, cis-polybutadiene, butadieneacrylonitrile rubber, Neoprene Rubber, butyl rubber, (BB), silicone rubber, polysulfide, urethane rubber, polyacrylate, propylene oxide rubber, fluorosilicone rubber, chlorosulfonated polyethylene, chlorinated polyethylene, fluorohydrocarbon rubber, and so on.

It is generally desirable that the two layers be bonded to each other. If the inner layer is the heat recoverable layer the external rubber layer may stick or adhere tightly enough to it by virtue of its tendency to contract down onto the heat recoverable layer. In cases for some other applications where the elastomeric material is the inner layer, a stronger bond may be desired. A bond can be developed by several means. Any suitable adhesive can be used to bond the heat recoverable and elastomeric layers together. Examples of such adhesives are laminating adhesives, such as polyesters, polyurethanes. Peroxides, either organic or silyl, which form crosslinks between the two layers; structural adhesives such as epoxies, nitrile rubber, phenolics,, hot melt adhesives of suitable bond strength and softening temperatures, such as polyamides and various rubber-based adhesives such as those based on silicone nitrile and neoprene [19].It may be desired to bond or weld both the layers together without using an adhesive. Such bonding may be achieved simply by heating the layers to a high enough temperature at their interface so that they become soft and flow able and then applying sufficient pressure to achieve a bonding or welding of the layers [20-22].

General applications of heat-shrinkable items are caps, sleeves, pipes, tubes, multi-way cable breakouts etc. In order to achieve the best quality the materials need to undergo certain specifications during the manufacturing process and application stages. Therefore there is also a great requirement to develop new materials tailored or manufactured exactly to specific applications. But developing such tailored products requires overcoming certain challenges like dispersibility, processability, exact mechanical properties (expandability, shrink ratio and temperature, flexibility, tensile properties, hardness), continuous operating temperature, chemical resistance, flame retardancy, electrical and dielectrical properties, and crack resistance. These are some of the criteria's that should be taken into consideration during the process of development of new materials for heat-shrinkable items [23]. As a result the process need several materials some times more than 10 ingredients and there is the possibility of unwanted interactions between them [24]. The aim of this paper was to focus on some selected topics of technological importance, related with the research and development of work dedicated to modern materials for production of heat-shrinkable items.

6.1. Application of encapsulated TPO's

- TPO heat shrink tubing is used to protect wires, for better abrasion resistance as well as environmental protection for solid wire conductors, connections, joints and terminals used in several industrial and domestic fields. It also has application to modify or repair the insulation on wires or to bundle them together, to protect wires or small parts from minor abrasion, offering environmental sealing protection. The tubing provides good electrical insulation protection from dust, solvents and other foreign materials, and is mechanically held in place by its tight fit [25-27].

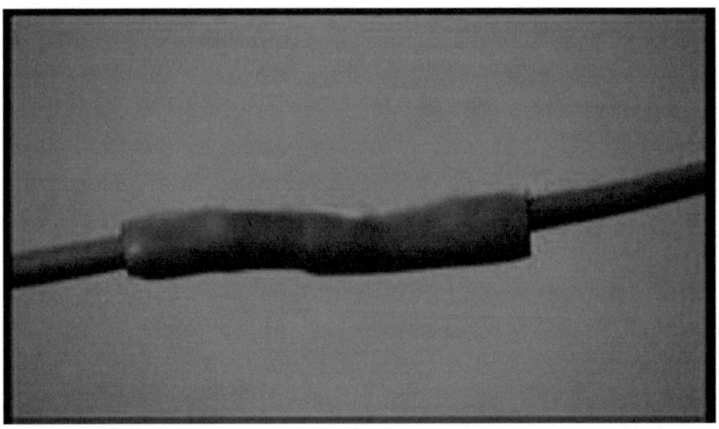

Sheme 4. TPO heat shrinkage tube

Some types of heat shrink tube contain a layer of thermoplastic adhesive on the inside to help provide a good seal and better adhesion, while others rely on friction between the closely conforming materials.

The manufacturing process of these types of tubes generally follows certain simple steps. The raw material is chosen based on its properties and end use application. The material are always manufactured along with many additives like certain stabilizers, colorants based on the application concerned. It can be briefly discussed for better understanding. At first the tube is

http://dx.doi.org/10.5772/61691

extruded from the raw or neat material. Then the tube is subjected to a separate process where it is cross-linked. The material is often cross-linked through the use of electron beams, peroxides, or moisture. The main objective for cross-linking is to create a memory in the tube so that it is able to shrink back to its original extruded dimensions upon heating. Then the tube is heated just above the polymer's crystalline melting point and expanded in diameter, often by placing it in a vacuum chamber [28-29]. After this step, in its expanded form a rapid cooling is applied. Later, when heated (above the crystalline melting point of the material), the tubing shrinks back to its original extruded size. For external use, heat shrink tubing often has a UV stabilizer added.

- PO tubes, are the most widely and commonly used tubes in some major field in military, aerospace industries. It has maximum continuous-use temperatures from -55 to 135 °C. The main advantage of using PO tubes are their flexibility and fast-shrinking property, and they are manufactured in a wide range of colors (including clear), which can be used for color-coding wires also. Black color PO tubes tend to have lower resistance to ultraviolet light; accordingly, and are only recommended for outdoor applications. A common shrink ratio is 2:1, but high-grade polyolefin heat shrink is also available with a 3:1 ratio [30].

There are several other materials that offers resistance to diesel and aviation fuels, and also some woven fabric, providing increased abrasion resistance in harsh environments.

Heat shrink tubing is also available in a wide variety of colors for color-coding of wires and connections. Since long heat shrink tubing has been used for PC to tidy up the interior of computers and provide an appearance considered pleasing. In response to this opening market manufacturers have started producing heat shrink tubing in luminous and UV reactive varieties.

Although heat shrink is usually used for insulation, heat shrink tubing with a conductive lining is available, for use particularly on joints that are not soldered. Heat shrink end caps, closed at one end, are used to insulate the exposed cut ends of insulated wires.

Sheme 5. Heat shrink insulated wires

Cross linked polyolefin heat shrink flame retardant sleeving with an adhesive lining. Designed to provide a permanent encapsulation for protection against moisture in a wide variety of applications once heated.Operating temperature -55°C to +125°C.(Source Google)

Sheme 6. Cross linked polyolefin heat shrink flame retardant sleeving

Heat shrinkable Polyolefin tubing with an integrally bonded adhesive inner lining, designed to provide a permanent encapsulation for protection against moisture in a wide variety of applications such as electrical wire splices, cable jackets, wire breakouts, strain reliefs and protection boots for electrical components. (Source Google)

7. Conclusion

Use of TPEs is increasing for wide range of applications. Encapsulation is one of the areas where TPEs are hugely used. About 85% of plastic products in the modern world are made up of TPEmaterials of which a huge percentage is occupied by TPO's as they provide comparatively high price performance as compared to other natural rubbers. The selection of a TPO material for any particular end application requires a special focus on its physical, chemical,mechanical and electrical properties. Polyolefin elastomers (TPO's) have proven their viability in flexible plastics applications and use in a variety of industries. Further advances in application development, product design, and manufacturing capabilities will provide increasing opportunities for years to come. TPO's has proven to be the best material used for outdoor encapsulation application. One of the major criteria for TPO's suitability in this sector is the heat sensing ability. There is a direct relation between the processing parameters and the shrinkability of the TPO polymer blends. With increase in crosslinking and the cure time the shrinkability increases. The shrinkability also depends on the curing agent taken into consideration. Addition of flame retardant decreases the stretchability of the material.

Author details

Tanya Das[1*] and Sunanda Roy[2]

*Address all correspondence to: tanya.das@bears-berkeley.sg

1 Berkeley Education Alliance for Research in Singapore Ltd, National University, Singapore

2 Materials Science and Engineering, Nanyang Technological University, Singapore

References

[1] Walker BM, Rader CP. Handbook of Thermoplastic Elastomers. Van Nostrand Publishers,Reinhold, NewYork,1988.

[2] Legge NR, Holden G, Schroeder HE. Thermoplastic Elastomers: a Comprehensive review. Carl Hanser Verlag,Munich, 1987; 80: 574.

[3] Laurer JH, Mulling JF, Khan SA, Spontak RJ, Bukovnik R. Thermoplastic elastomer gels. I. Effects of composition and processing on morphology and gel behavior. J Polym Sci Part B: Polym Physics. 1998; 36: 2379–2391

[4] Pramanik M, Srivastava SK, Samantaray BK, Bhowmick AK. Synthesis and characterization of organosoluble, thermoplastic elastomer/clay nanocomposites. J Polym Sci Part B: Polym Physics. 2002; 40: 2065–2072

[5] Barraza HJ, Pompeo F, O'Rea EA, Resasco DE. SWNT-filled thermoplastic and elastomeric composites prepared by miniemulsion polymerization. Nano Lett. 2002; 2: 797–802

[6] Oksman K, Lindberg H. Influence of thermoplastic elastomers on adhesion in polyethylene–wood flour composites. J Appl Polym Sci. 1998; 68: 1845–1855

[7] Qingle.Z, Yongxin P, Demin J.Study on toughening of up with liquid rubbers. J. Polym Mater Sci & Engin. 1998; 5: 5.

[8] Kear KE. Developments in Thermoplastic Elastomers. Chemtec Publishing. 2003

[9] Amin M, Salman M. Aging of polymeric insulators (an overview). Rev. Adv. Mater. Sci. 2006;13: 93-116

[10] Gardner RJ, Martin JR. Effect of relative humidity on the mechanical properties of poly(1,4-butylene terephthalate). J Appl Polym Sci. 2006; 25: 2353-2361

[11] Lievana E, Karger-Kocsis J. Use of Ground Tyre Rubber (GTR) in thermoplastic polyolefin elastomer compositions. Progr Rub, Plast Recy Technol, 2004; 20:1

[12] Abdou-Sabet S, Puydak RC, Rader CP. Dynamically vulcanized thermoplastic elastomers. Rub Chem and Technol. 1996; 69: 476-494

[13] Rieth LR, Eaton RF,Geoffrey W. Polymerization of ureidopyrimidinone-functionalized olefins by using late-transition metal Ziegler–Natta catalysts: Synthesis of thermoplastic elastomeric polyolefins. Angewandte Chemie Int Ed. 2001; 40: 2153-2156

[14] Vocke C, Anttila U, Heino M, Hietaoja P, Seppälä J. Use of oxazoline functionalized polyolefins and elastomers as compatibilizers for thermoplastic blends. J Appl Polym Sci.1998; 70: 1923-1930

[15] Mishra JK, Raychowdhury S. Heat shrinkable polymer blends based on grafted low density polyethylene and chlorosulfonated polyethylene. Intern. J. Polym Mater., 2000; 47: 407-421

[16] Tjong SC, Ruan YH. Fracture behavior of thermoplastic polyolefin/clay nanocomposites. J Appl Polym Sci. 2008; 110: 864-871

[17] Yuechun M, Farinha JPS, Mitchell A. Winnik, Yaneff PV, Ryntz RA. Compatibility of chlorinated polyolefin with the components of thermoplastic polyolefin: A study by laser scanning confocal fluorescence microscopy. Macromolecules. 2004; 37: 6544–6552

[18] Zheng Y, Yanful EK, Bassi AS.A review of plastic waste biodegradation. Critical Rev Biotechnol. 2005; 25: 243-250

[19] Chaoqun Li, Hua Deng*, Ke Wang, Qin Zhang, Feng Chen andQiang Fu. Strengthening and toughening of thermoplastic polyolefin elastomer using polypropylene-grafted multiwalled carbon nanotubes. J Appl Polym Sci. August 15, 2011, 121, 4, 2104–2112,

[20] Wong S, Lee JWS, Naguib HE, Park CB. Effect of processing parameters on the mechanical properties of injection molded thermoplastic polyolefin (TPO) cellular foams. Macromol Mater and Engin. 2008; 293: 605-613

[21] Wong TL, Barry CMF, Orroth SA.The effects of filler size on the properties of thermoplastic polyolefin blends. J Vinyl Add Technol. 1999; 5: 235-240

[22] Bessede JL, Huet I, Kieffel Y. Suitability of thermoplastic for the making of HV gas insulated substation insulator. Conference on Electrical Insulation and Dielectric Phenomena. 2005; 42: 12-15

[23] Lu J, G Wei G, H.-J. Sue H, Chu J. Toughening mechanisms in commercial thermoplastic polyolefin blends. J Appl Polym Sci.2000; 76: 311-319

[24] [24].Kashif S, Imdad, Amin M. Aging of polymeric insulator installed at colombian environment. Int J Engin & Comp Sci IJECS-IJENS 2006; 12: 51-54

[25] [25] Amin M, Amin S, Ali M. Monitoring of leakage current for composite insulators and electrical devices. Rev. Adv. Mater. Sci 2006; 21: 75-89

[26] Chattopadhyay S, Chaki TK, Bhowmick AK. Heat shrinkability of electron-beam-modified thermoplastic elastomeric films from blends of ethylene-vinylacetate co-polymer and polyethylene. Radiat Physics Chem. 2000; 59:501-510

[27] Lopez MA, Burillo G, Charles A.Studies on the memory effect in polyethylene. Radiat. Phys. Chem. 1994; 43: 227-231.

[28] Kenaeth G, Budinski. Resistance to particle abrasion of selected plastics. Wear.1997: 302-309

[29] Abdou-Sabet S, Puydak RC, and Rader CP. Dynamically vulcanized thermoplastic elastomers. Rub Chem Technol: July 1996, 69,. 3, 476-494.

[30] Böhm H, Betz S, Ball A. The wear resistance of polymers. Tribol Int. 1990; 23:399–406.

Chapter 6

Thermoplastic Elastomers with Photo-actuating Properties

Markéta Ilčíková, Miroslav Mrlík and Jaroslav Mosnáček

Additional information is available at the end of the chapter

http://dx.doi.org/10.5772/59647

Abstract

This contribution reviews elastomeric materials with photo-actuation behavior with emphasis on thermoplastic elastomers and their composites. The principles of the photo-actuation and the main factors affecting the photo-actuation phenomena of thermoplastic elastomer materials are discussed in detail. The well-performing photo-actuating systems involving both statistical and block copolymers-based thermoplastic elastomers are assessed in terms of their advantages and limitations. Methods for evaluation of photo-actuation behavior of the materials are reported as well. Finally, the utilization of the photo-actuating thermoplastic elastomers is presented.

Keywords: Thermoplastic polyurethane, styrene-isoprene-styrene, carbon nanotubes, graphene, liquid crystals

1. Introduction

MERGEFORMAT Actuation phenomenon is considered as a material's ability to undergo reversible shape changes in response to an external stimulus [1–3]. There are several trigger stimuli reported, such as the electric field, light, pH, or temperature [4–8]. The trigger-responsive materials find their employment in a wide range of applications comprising sensor, artificial muscles, etc. [1, 2]. Photo-induced actuation technologies can offer many advantages over traditional, mainly electrically driven, actuators, such as remote energy transfer, remote controllability, better scalability, low electromagnetic noise, easy construction, and capability of working in harsh environments [7]. Generally, the actuating materials can be pure polymers or polymer composites. In both cases, the energy absorber-triggers and assembling structures need to be present in the materials [2]. However, the actuation may be improved by the addition of an energy trigger. As an energy absorbers, dyes [9] or carbon based fillers were reported [10, 11].

Various elastomers were investigated for their photo-actuation behavior, including liquid crystalline elastomers, poly(dimethylsiloxane), various thermoplastic elastomers (TPEs) based on polyurethane, poly(ethylene–*co*-vinyl acetate) (EVA), polystyrene-*block*-polyisoprene-*block*-polystyrene (SIS), and acrylic-based block copolymers, as well as hydrophilic copolymers such as NAFION or hydrogels based on copolymers of acrylic acid and *N*-isopropyl acrylamide. The principles of photo-actuation depend on the type of elastomer and type of the light absorbers.

The most common principle of photo-actuation of chemically cross-linked elastomers and/or TPEs is based on the presence of soft segments responsible for shape changes under illumination, and hard segments responsible for returning the material to its stage before illumination [12]. Thus, the pre-strained material containing some light absorbers absorbs the energy from the light and converts it to heat that is conducted through the material. The heat causes that the pre-strained polymer chains in the soft segments shrink, i.e., contract to form coil, and that results in shape changes of the material (Figure 1). The hard segments are formed from chemically or physically cross-linked parts, enabling the reversibility of the actuation by the energy balance between absorbed and released energy in the form of mechanical response. Therefore, sometimes the term photo-mechanical actuation is also used for the photo-actuation phenomenon [13].

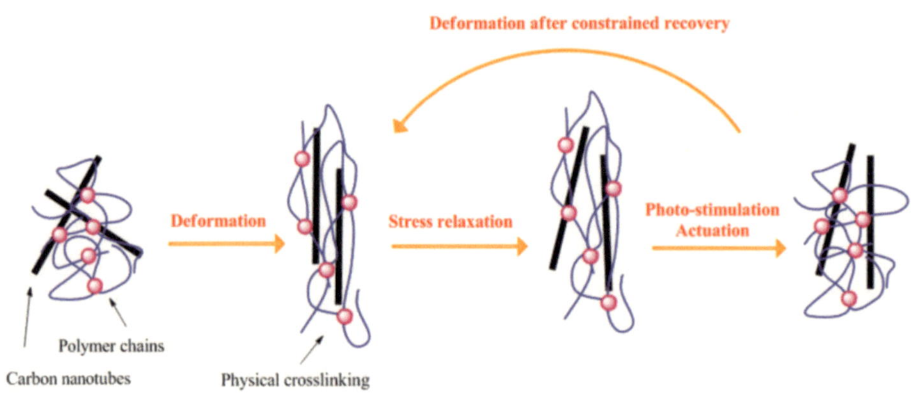

Figure 1. Principle of photo-actuation behavior of physically or chemically cross-linked systems.

Thermoplastic polyurethanes (TPUs) materials have unique basis of the photo-actuation phenomenon. Depending on the chemical composition, the polymer chains consist of soft segments with melting point ranging from 35°C to 50°C and hard segments with melting point exceeding 100°C. During application of certain pre-strain to the material, the soft polymer

http://dx.doi.org/10.5772/59647

chains are crystallized (Figure 2). These crystallites provide additional physical cross-linking that stabilize the polymer chains in elongated state. The actuation occurs after melting the soft crystallized segments as a recoiling of the soft polymer segments. In case of light induced actuation, the energy from the light source transported to the polymer chains must be high enough to reach the melting point of the soft segments [12].

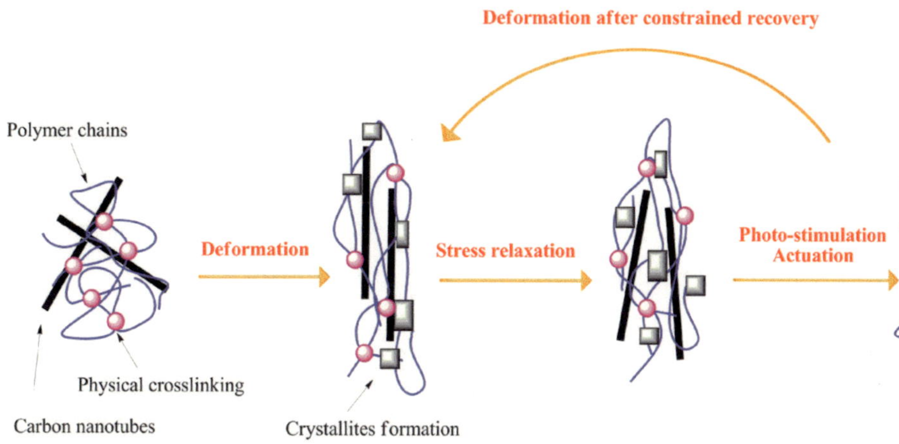

Figure 2. Principle of photo-actuation behavior of thermoplastic polyurethanes [12].

The main principle of the photo-actuation of liquid crystals and liquid crystalline elastomers is based on the ordering and disordering of the structure upon light stimulation (Figure 3) [14–16]. The material is able to be reversibly transformed between these two phases as a result of stimulation, while actuation stress is created. The magnitude of the shape changes and the stress development are strongly influenced by various factors, such as toughness of the polymer matrix, modification of the polymer matrix, or type of additive.

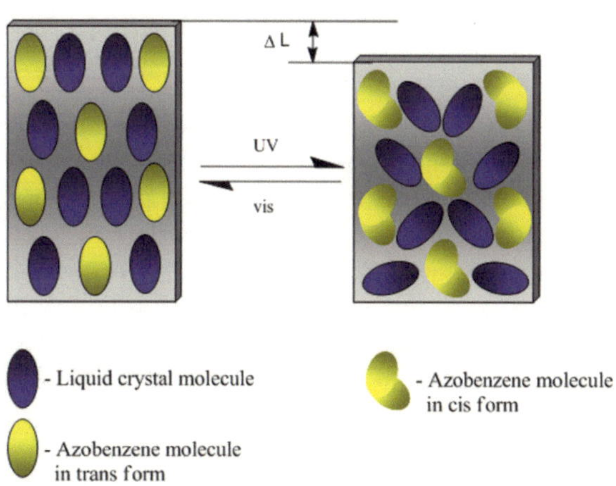

Figure 3. Principle of photo-actuation behavior of liquid crystals systems [14].

Very rare photo-actuating principle was found on the hydrogel samples [17]. In this case the actuation principle is based on the difference of swelling in the dark and upon the light and will be described more in detail later. Such changing of dark/light conditions is not that fast as in case of previous principles; however, the potential applications of such systems are very promising and the preparation of the new materials with shorter response time is highly challenging.

Here the photo-actuation of various systems and their advantages and disadvantages will be reviewed, while photo-actuating systems based on thermoplastic elastomers will be described more in detail. Analytical methods used for determination of photo-actuation behavior of various systems will be discussed as well. Finally the applicability of the photo-actuating TPE will be referred.

2. Photo-actuation of chemically cross-linked elastomers

2.1. Liquid crystal elastomers

Liquid crystals (LCs) are the most common materials frequently applied in order to provide the system with good actuation behavior [18]. This unique property is allowed due to the structure of the LCs that combine the mobility of the isotropic liquids and orientation order of crystalline solid [15]. By incorporation of the light-triggered materials to the LCs, the alignment can be properly controlled over large areas, and thus the materials can be effectively utilized in photonic applications such as signal processing, optical switching, or already mentioned photo-actuation [19–21].

However, it was found out that photo-chemical phase transitions disrupt the LC phase and the material turns to isotropic [16]. Thus, the research interest was further focused on the preparation of the cross-linked LCs structures. Since Finkelmann et al. showed the photo-contraction of liquid crystal elastomers (LCEs) and Ikeda provided the system with photo-actuating LCE films [22], the main aim of the scientific groups has been focused on the preparation of stable and well photo-actuating systems.

The main principle of LCEs utilization is the weak cross-linking of ordered macroscopical structure, which provides the benefits of orientation order of liquid crystals and elasticity of common rubbers [23, 24]. The typical case of cross-linked liquid crystalline network is LCE in the form of uniaxially oriented planar film where the gradient in light intensity through the thickness of the film causes photo-actuation [16, 25–28]. Unfortunately, the response to the external stimulus (light) is rather slow and shape changes are small due to the low thermal conductivity resulting in low energy transfer within the material. This drawback can be solved by the incorporation of light-triggered materials similarly as in the case of LCs, or by addition of fillers improving the thermal diffusivity within the whole final material.

In order to perform the LCEs with excellent photo-mechanically responsive properties, light-sensitive monomer can be used. The final materials containing mainly azobenzene derivatives belong to highly photo-actuating materials [29–31]. This material is unique due to its dynamics of *trans* to *cis* isomerization of azobenzene unit resulting in the change of absorption spectrum able to report the local rigidity of its surroundings [32]. Thus in ordered liquid crystalline network, trans-cis isomerization leads to order reduction resulting in the macroscopic contraction in the main-chain direction and an expansion in the opposite direction. Such systems exhibit considerably improved photo-actuating properties [33].

Another additive, which can considerably improve the photo-actuating properties of the LCEs, are carbon nanotubes (CNTs). CNTs are efficiently applied due to their one-dimensional shape, nanoscale diameters, large surface area, and excellent electrical and thermal properties [34, 35]. Special case of carbon nanotubes effectively applied in photo-actuating systems are single-walled carbon nanotubes (SWCNTs) exhibiting strong absorptions in the visible and near-IR region [36]. Therefore, they efficiently convert the light energy into thermal energy providing the heat source at nanoscale and thus create the thermal pathways within the liquid crystal elastomer upon IR irradiation [37, 38]. However, the utilization of the CNTs in the case of LCEs is rather limited since the proper dispersion of CNTs in case of higher loading (above 1 wt. %) is very difficult especially without surface modification [38, 39]. This drawback can be solved by incorporation of the pyrene to both ends of the LCE chains [40]. π-π interactions between the pyrene group and CNTs can significantly improve the dispersion of the CNTs in LCEs [38].

Finally, it can be concluded that from the group of liquid crystals, mainly liquid crystal elastomers are effectively applied due to their excellent physical properties. However, in order to facilitate the photo-mechanical response of those materials, various additives can be efficiently utilized. The additives, such as azobenzene-based derivatives or carbon nanotubes provide better absorption of the light and improve thermal conductivity of the materials.

2.2. Other chemically cross-linked elastomers

Poly(dimethyl siloxane) (PDMS), which was often used as a main chain LCE, was effectively used as a matrix also in other systems for photo-actuation applications. This versatile material have many potential applicability in the medical field, due to good biocompatibility, low glass transition temperature, and linear elasticity over broad temperature range (-50°C–200°C) and strains [41]. Solely, PDMS has very poor mechanical response to the light because of low thermal conductivity. Hence, PDMS matrix has to be filled with carbon-based fillers. The composites based on graphene-nanoplatelet were found to exhibit improved thermal conductivity of the samples, and thus also enhanced photo-actuation response [11, 42]. The contraction of the irradiated samples was obtained only above the certain pre-strain (above 10 %). In the case of composites containing thermally reduced graphene oxide, the pre-strain of 9% was sufficient to observe the contraction [7]. Nearly twice higher photo-actuation stress was obtained in comparison with CNTs-containing PDMS composite materials [1, 43, 44]. Thus, PDMS-based materials are very promising for their application as a photo-actuator mainly after the incorporation of the light-triggered fillers such as CNTs or graphene oxide. Those fillers significantly improve the thermal conductivity, resulting in better heat exchange within the material and thus provide the systems with promising photo-actuation performance.

Special photo-actuating materials are based on hydrogels [17]. The hydrogels were made by copolymerization of *N*-isopropyl acrylamide with various ratios of acrylic acid (AA). Benzospiropyran was used as an energy absorber in this system. The swelling of the samples in water changed with switching on/off the light. When the samples were exposed to the light the relative gel swelling was 78 %, while after switching off the light, the relative gel swelling increased up to 96%. This photo-actuation behavior was stable up to five light-dark cycles. This unique property is mainly caused by the utilization of the AA enhancing the proton transfer within the sample when exposed to light or dark and improving the swelling/ contraction behavior (Figure 4).

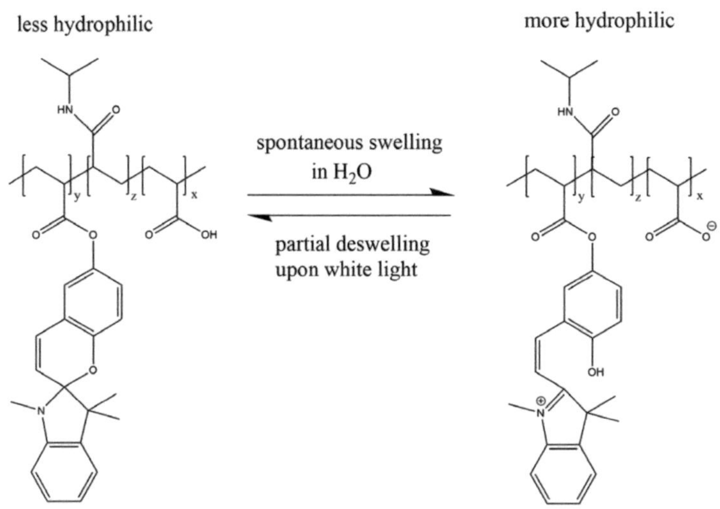

Figure 4. Schematic representation of the proton exchange taking place in hydrogels between the acrylic acid and the benzospiropyran in the dark and under irradiation [17].

3. Photo-actuation of thermoplastic elastomers

TPEs having physically cross-linked structure possess several advantages compared with chemically cross-linked elastomers. The main advantages are a repeatable processability of TPEs and possibility of tuning their mechanical properties by choosing the different segments to tailor finely the required properties. TPEs are also relatively cheap compared with liquid crystal-based systems. Therefore, the TPE systems are very promising not only from laboratory or specified purposes point of view, but also in terms of large-scale industrial application.

All of the TPE photo-actuating systems utilize light-triggered additives, mainly carbon-based particles, in order to enhance their photo-actuation response [45–47]. However, the poor dispersibility of the additives is a common problem. If the light-trigged additive is not well dispersed, the photo-actuation response will be of low performance. Nevertheless, this drawback can be effectively solved by suitable surface modification either covalent [48] or non-covalent [49].

3.1. Poly(ethylene-*co*-vinyl acetate)

One of the statistical copolymers studied for its photo-actuation properties is poly(ethylene-*co*-vinyl acetate) (EVA). The ethylene units in EVA provide for semi-crystalline properties of the copolymer, while the vinyl acetate units form amorphous part. The copolymerization of ethylene with vinyl acetate enhances the crystallization of the ethylene units and the crystallization degree of ethylene units increases with increasing amount of vinyl acetate in the copolymer. The final copolymer structure includes two segments: hard segment consisting of ethylene crystalline phase and soft segments represented by amorphous vinyl acetate and amorphous ethylene phase. Thanks to this unique behavior, EVA copolymers provide the materials with tunable elasticity depending on the vinyl acetate content.

The photo-actuation phenomenon has been studied on two EVA copolymers, EVATAN and LEVAPREN 500, with different content of vinyl acetate (28% and 50%, respectively) [47, 50, 51]. Generally, the selection of EVA with appropriate content of vinyl acetate can be crucial. EVA with too low vinyl acetate content will have lower melting point and loss of elastic properties can occur during photo-actuation cycles. The reason is the possible melting of the crystalline phase due to local overheating of the material after absorption of the light by carbon fillers and the release of the energy to the material in the form of the heat. On the other hand, EVA with too high vinyl acetate content can be too tough to provide good photo-actuation.

The effect of the different type of carbon-based fillers such as MWCNTs and SWCNTs and their content in the EVATAN matrix has been investigated. In order to improve the dispersity of the CNTs within the matrix the surface of CNTs was non-covalently modified with cholesteryl l-pyrenecarboxylate (PyChol), based on π-π interactions between CNT surface and pyrene (Figure 5). The EVA composite containing PyChol-modified MWCNTs exhibited higher values of photo-actuation stress compared with PyChol-modified SWCNTs (Figure 5). Better light absorption of the MWCNTs enhancing the photo-mechanical response of the EVA copolymer was suggested as a possible explanation. It has also been found that MWCNT/EVA systems with 0.1 wt. % of PyChol-modified MWCNTs exhibited higher photo-actuation stress compared with the system containing 3 wt. % of PyChol-modified SWCNTs. Even though the authors did not comment on this result, the possible explanation could be that the elasticity of the CNTs/EVA systems is decreased at higher content of the filler, thus suppressing the possibility of this material actuate effectively upon light stimulation.

Similarly, incorporation of 0.1 wt. % of PyChol-modified MWCNTs into the LEVAPREN matrix provided well-dispersed systems with appropriate photo-actuation behavior upon light stimulus [51]. This study was mainly focused on the possibility of the material to provide the system with repeatable photo-actuation phenomenon. In addition the response of this material on various light intensities has been investigated. It has been shown that increasing the intensity of the light led to more pronounced changes but with slower response on the switching on/off the light.

Different methods were used for investigation of photo-actuation behavior of the systems based on EVATAN and LEVAPREN matrices; therefore, the effect of EVA matrix composition cannot be directly compared. Generally, it has been shown that both investigated EVA matrices

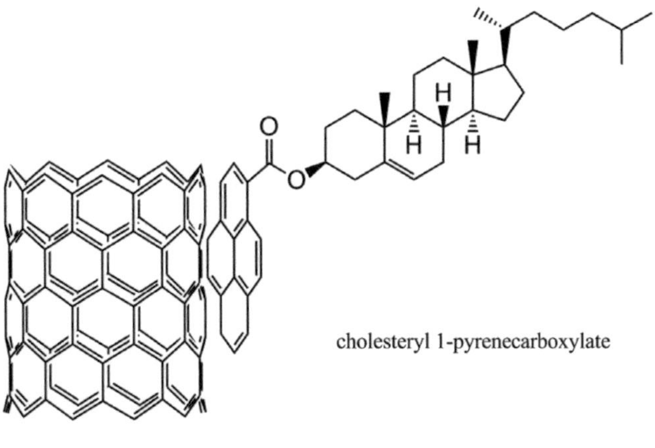

Figure 5. Schematic illustration of non-covalent modification of CNT with cholesteryl 1-pyrenecarboxylate [50].

can be utilized for photo-actuation applications, while at least for LEVAPREN, i.e., matrix with higher melting point (T_m = 96°C and 71°C for LEVAPREN and EVATAN, respectively), also repeatable photo-actuation behavior has been proved.

3.2. NAFION

Perfluorosulfonated ionomer NAFION is another statistical copolymer with properties of TPEs that has been investigated for its applicability as photo-actuating material [52, 53]. NAFION solely exhibits only poor ability of photo to mechanical energy conversion. Hence, SWCNTs were used as the light-triggered material for preparation of SWCNTs/NAFION bilayer samples providing the good photo to mechanical energy conversion. The determined actuation changes in light switching on/off cycles were 200 μm and 600 μm at light intensity of 18 mW cm^{-2} and 75 mW cm^{-2}, respectively. In SWCNTs/NAFION bilayer system, the photo-actuation phenomenon is obtained by redistribution of the hydrogen ions and water molecules in the SWCNTs/NAFION interfacial region. The light establishes an electric field at the interface that promotes the hydrated hydrogen ions toward the interface. Both the SWCNTs and NAFION contribute to the mechanical action, because water molecules are depleted in the pores of NAFION near the interface and they occupy the interior of the SWCNTs leading to improved effect of contraction of the interfacial region between the SWCNTs and NAFION (Figure 6) [52].

3.3. Thermoplastic polyurethanes

TPUs are special type of copolymers, where the hard and soft segments are repeatedly alternating. The hard polymer segments are represented by diisocyanates [54], such as 4,4'-diphenylmethane diisocyanate (MDI), hexamethylene diisocyanate (HDI), 3,3'-dimethyl-4-4'-biphenyl diisocynate (TODI), etc. On the other hand, the soft segments are mainly represented by polyesters [55] or polyethers [56]. However, the real composition of the TPUs usually also includes chain extenders [57] as can be seen in Figure 7. The chain extender is mainly based

http://dx.doi.org/10.5772/59647

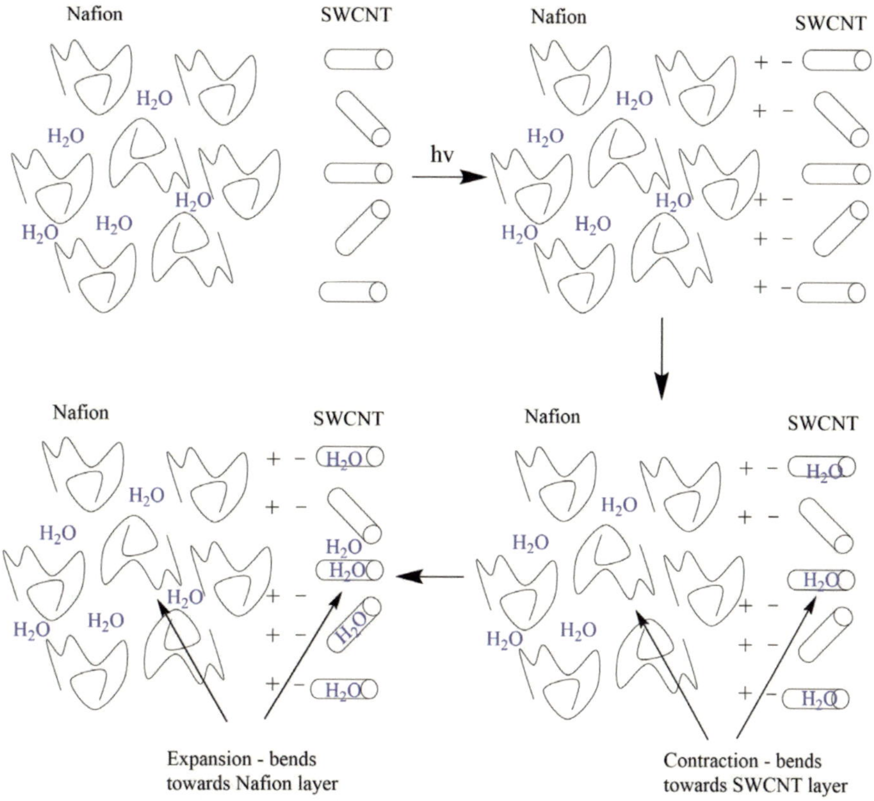

Figure 6. Photo-actuation mechanism of the Nafion-SWCNT bilayer upon light stimulation [52].

on linear diols, such as ethylene glycol, 1,4 butadiene, 1,6 hexandiol, etc. This unique composition of the TPUs enable tunability of the mechanical properties and therefore TPUs are very promising material for the photo-mechanical actuation.

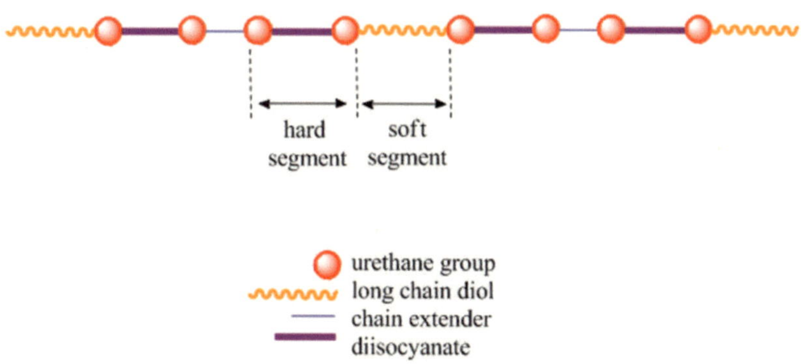

Figure 7. Composition of the thermoplastic polyurethane elastomers containing chain extenders.

Similar to the previous matrices, TPUs solely exhibit only moderate photo-actuation performance. This is mainly due to poor strain-induced crystallization of the soft segments upon tensile deformation. Hence, all research groups have focused their investigations on the effect of the fillers addition on the strain-induced crystallization of the soft segments upon tension and its connection with the photo-actuation performance of the TPUs composites [12, 58–60]. An impact of the CNTs (5 wt. %) and carbon black (5–20 wt. %) on the mechanical properties, strain-induced crystallization and final photo-actuation performance of TPU (Irogran PS455-203) composites has been investigated (Figure 8) [12]. For the CNT reinforced matrix the young modulus increased by factor 2–5 and considerably improved strain-induced crystallization was observed by DSC and proved by XRD measurements upon tension. Recovery of TPUs composite containing 5 wt. % carbon black, actuated by infrared (IR), was only 25%–30% of the stress achieved by heat actuation, compared to the almost 100% for CNTs containing TPUs composites. Four times higher amount of carbon black (20 wt. %) compared with CNTs (5 wt. %) was needed to obtain comparable deformation of TPU samples under IR irradiation.

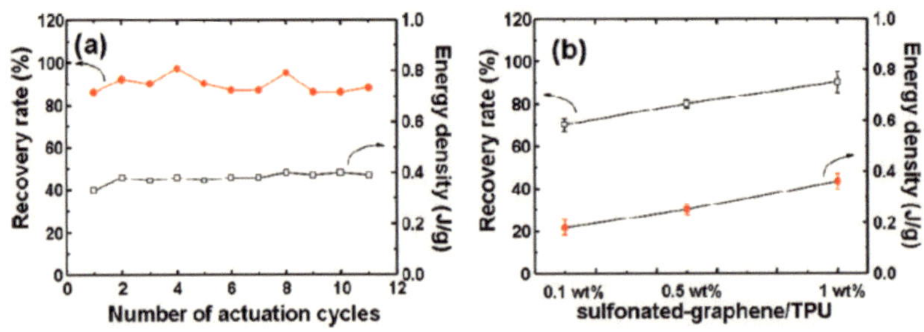

Figure 8. The recovery rate for the samples containing SRGO particles. Reprinted with permission from Liang JJ, Xu YF, Huang Y, Zhang L, Wang Y, Ma YF, et al. Infrared-Triggered Actuators from Graphene-Based Nanocomposites. Journal of Physical Chemistry C. 2009;113(22):9921-7. Copyright © 2009 American Chemical Society [58].

It has been reported that well-dispersed graphene in MDI-based TPUs provided the system with enhanced strain-induced crystallization, while the poor dispersity of the same graphene in HDI-based TPUs led to the suppressed strain-induced crystallization [59]. Better dispersity of graphene in MDI-based TPUs compared with HDI-based TPUs was attributed to π-π interactions between graphene and aromatic ring of MDI. Further improvement of compatibility between graphene and TPUs was obtained after surface modification of graphene. Increasing amount of hydroxyl groups on graphene surface by modification with methanol allowed incorporation of graphene in TPUs structure [59, 61, 62]. Such grafting of TPUs on graphene improved shape recovery (Figure 8). Contrary that it has been shown that intimately mixed graphene disturbed hydrogen bonds between hard segments of SPU balancing the reinforcing effect of graphene. The photo-actuation phenomenon of these systems has been investigated upon IR irradiation and the shape recovery after IR stimulation ranged 90%–99% of deformation.

Influence of the sulfonation of reduced graphene oxide (SRGO) (1 wt. %) on mechanical, energy-transfer and photo-actuation performance of TPU-based (Irogran PS455-203) composites has been investigated as well [58]. The composite sample containing SRGO exhibited highest absorption in IR region (500 nm - 1000 nm) compared to neat TPU matrix and composites containing 1 wt. % of isocyanated-graphene oxide and reduced graphene oxide. The reason is considerably restored sp^2 network in SRGO. Due to the increased IR absorption, the sulfonated SRGO/TPUs composites contracted faster than others and also the recovery rate was 15% faster compared with isocyanate-graphene oxide or reduced graphene oxide composites (Figure 9).

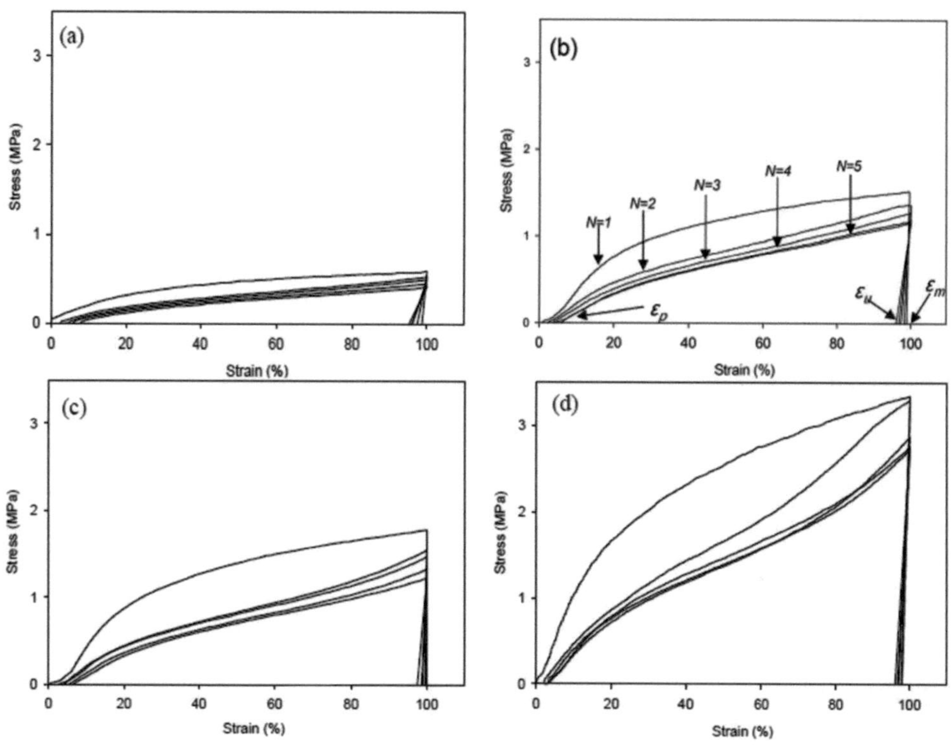

Figure 9. Photo-actuation behavior of the sample (a) pure TPU, (b) TPU with 5 wt. % GO, (c) TPU with 10 wt. % GO, and (d) TPU with 20 wt. % GO [59].

In order to obtain high actuation performance, structural uniformity, good dispersion, and high purity of carbon fillers is crucial. Properties of the carbon fillers are affected by surface defects, large bundles, impurities, the anisotropy, and a structural mixture [63]. Surface functionalization of graphene in order to obtain their good dispersion in polymer matrix results in the decrease in thermal conductivity due to structural defects [64, 65]. Therefore, a balance between the restoration of sp^2 network and the dispersion of graphene is crucial for sufficient reinforcement and high thermal conductivity. Several studies have shown enhanced thermal conductivity and mechanical strength of polymer composites when hybrid graphene/CNT nanofiller were used. The synergistic effect between well-dispersed graphene and CNTs can be tuned by weight ratio of graphene to CNT [66–70]. Expected enhancing of thermal conductivity led to the investigation of the influence of incorporation of various contents of sulfonated CNTs into the SRGO/TPUs composites on photo-actuation performance of the prepared composites [60]. IR absorption of sulfonated CNT/SRGO/TPUs composites was higher than SRGO/TPUs. The DSC results showed that the melting range of soft segment crystallites was shifted to higher temperatures with incorporation of the fillers. This temperature increase was ascribed to the heterogeneous nucleating effect of SRGO and CNT. As expected, all sulfonated CNT/SRGO/TPUs composites exhibited enhanced thermal conductivity. TPUs composite with sulfonated CNT/SRGO ratio of 1/3 showed the best IR-actuated stress recovery of lifting in 18 s. Remarkable IR-actuated recovery delivered the mechanical stress of 1.2 MPa assigned to thermal conductivity of 1.473 W/mK and Young's modulus of 23.4 MPa. It has been concluded that a trade-off between the stiffness and efficient heat transfer, which can be controlled by synergistic effect between SRGO and SCNT, is critical for high mechanical power output of IR triggered actuators. Therefore, SRGO/SCNT/TPUs composites combining high output forces, and good cycling stability are highly suitable for development of advanced photo-actuating systems.

3.4. Polystyrene-based block copolymers

The materials investigating for effective respond to the photo-stimulation include also A-B-A triblock-based TPEs. They consist of one polymer block with low glass transition temperature (T_g) providing the actuation of the material and two blocks with high T_g providing the elasticity and reversible shape change after switching off the light stimulus. Mechanical, and thus also the photo-actuation properties of A-B-A triblock-based TPEs can be finely tuned by choice of the monomers structure and ratio between the soft and hard blocks. Similarly, as in all previous cases, the light-triggered materials have to be added to provide the system with sufficient actuation performance. The main A-B-A triblock-based TPE investigated for photo-actuation is polystyrene-*block*-polyisoprene-*block*-polystyrene (SIS) triblock copolymer [39, 46, 71].

A reduced graphene oxide (rGO) has been successfully used as a light-triggered material in the SIS matrix [71]. Investigation of the effect of the filler content on the photo-actuation showed that the best performance was obtained for the nanocomposites containing 1.5 wt. % of rGO. The higher rGO content resulted in decreased response that was in consistence with change in the mechanical properties. Apparently, a creep has been observed during irradiation that is understandable with respect to extremely large pre-strain (up to 150%) and also due to the utilization of light source with 22 W cm^{-2} intensity, which could significantly contribute to the mentioned creep behavior.

Besides rGO, the MWCNTs have been investigated in SIS matrix as well. Both elongation and contraction were observed under irradiation depending on the applied pre-strain (Figure 10). Under light intensity of 1.5 W cm^{-2}, minimal 20% pre-strain was needed to obtain 0.2% contraction of the composite [39]. The highest actuation was obtained at 40% pre-strain with 1.1% contraction. It should be mentioned that the MWCNTs were used in very low concentration (0.01 wt. %). Too high loading of neat MWCNTs disturbs physical cross-linking due to preferential interactions of MWCNTs with polystyrene blocks.

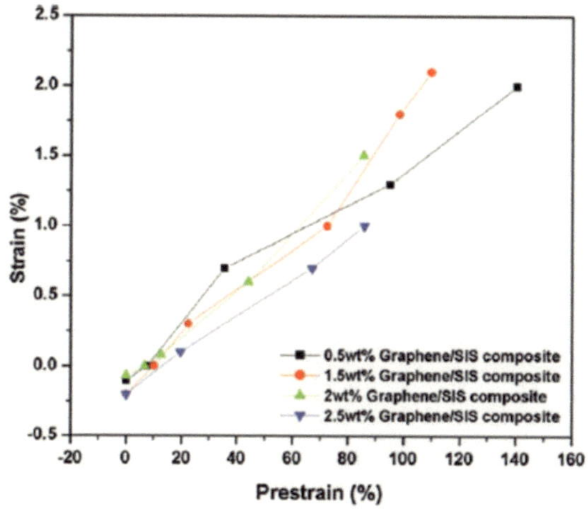

Figure 10. Photoactuation behavior of SIS samples with various contents of graphene particles. Reprinted with permission from Ansari S, Neelanchery MM, Ushus D. Graphene/Poly(styrene-b-isoprene-b-styrene) Nanocomposite Optical Actuators. Journal of Applied Polymer Science. 2013;130(6):3902-8. Copyright © 2013 Wiley Periodicals, Inc. [71].

In order to prevent the negative interactions of CNTs with SIS matrix resulted in deteriorating of the mechanical properties, the surface of CNTs was modified. Complex investigation of the structural, mechanical, and photo-actuation performance of CNT/SIS composites with both neat MWCNTs and covalently modified MWCNTs has been reported for filler loading of 0.15 wt. % to 3 wt. % [46, 72]. In order to tailor preferential interactions of MWCNTs with individual blocks of SIS matrix, the surface of MWCNTs was modified either with cholesteryl groups or with short polystyrene chains (Figure 11). In the former case the preferential interactions with polyisoprene phase, and in later case the preferential interactions with polystyrene phase were expected. The dynamic mechanical analysis (DMA) in wide temperature range (from -100°C to 150°C) has been used to investigate the specific interactions of selectively modified carbon nanotubes with individual blocks of SIS. The shift in T_g and activation energy of glass transition was compared. Addition of neat MWCNT resulted in the shift of T_g of both polyisoprene and polystyrene phase to lower temperatures, while the influence was more pronounced for polystyrene phase. The cholesteryl-modified MWCNT shifted the T_g of both polyisoprene and polystyrene phases to higher temperatures, while the shift for polyisoprene phase was more significant. The highest shift in T_g of polystyrene phase was observed in composites containing polystyrene-modified MWCNTs. Contrary to neat MWCNTs, in the case of polystyrene-modified MWCNTs the T_g shifted to higher temperatures. The activation energies of glass transitions followed a similar trend (see Table 1). In all investigated composites, an increase of MWCNTs concentration from 1 wt. % to 3 wt. % led to the deterioration of the mechanical properties. The highest actuation stresses were generated by composite containing low content of neat MWCNTs. However, the photo-actuation was irregular and fast drop of the baseline was observed as a result of negative interactions of neat MWCNTs with polystyrene phase and loss of the elastic properties. The most stable and reversible response was obtained in composite containing polystyrene-modified MWCNTs (Figure 12). In this case, however, the actuation stresses were only half of the stresses obtained for composites containing either neat

MWCNTs or cholesteryl-modified MWCNTs [72]. The reason is not effective energy transfer from polystyrene-modified MWCNTs, preferentially localized in hard polystyrene phase, to soft polyisoprene phase responsible for actuation changes. These extensive studies showed that the best photo-actuation could be expected in the case of selective localization of MWCNT in the soft phase of triblock thermoplastic elastomers with effective energy transfer only to soft phase polymer chains.

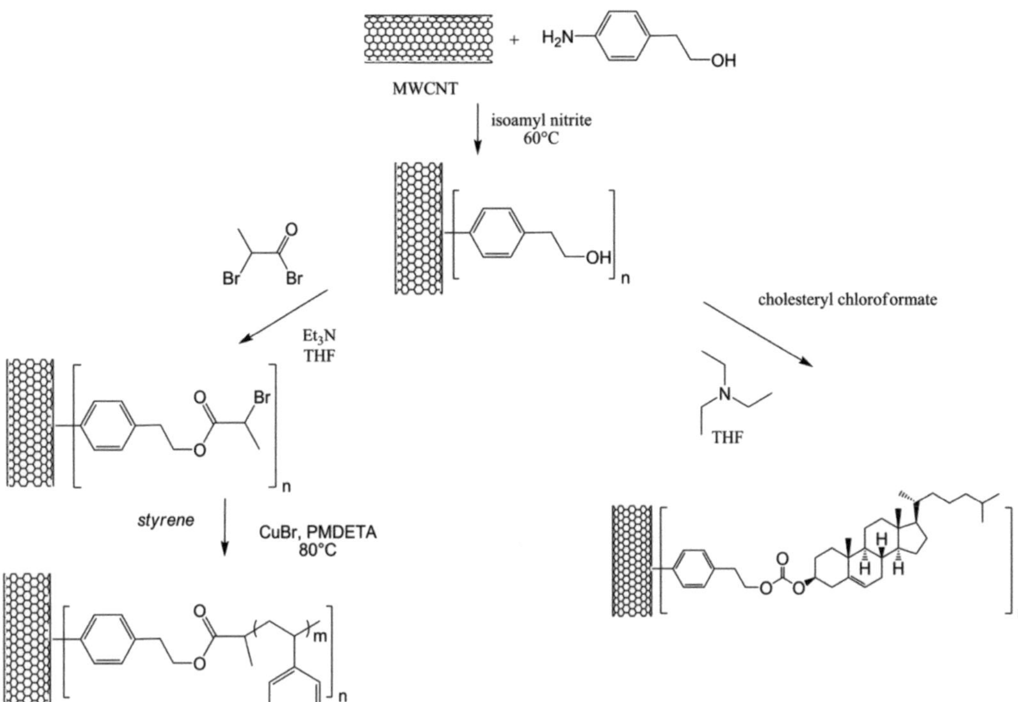

Figure 11. Covalent modification of multiwalled carbon nanotubes either by cholesteryl groups or with polystyrene chains [72].

In order to investigate the MWCNTs/SIS composites for their applicability in tactile displays, the Braille-like elements were prepared by thermoforming and their photo-actuation behavior has been investigated [46, 47]. In all experiments, blister expansion was observed as a result of low pre-strain induced during the thermoforming process. Regardless, the actuating response increased with intensity of the light source. The fastest response was observed in the composites containing polystyrene-grafted carbon nanotubes.

Another polystyrene-based triblock copolymer investigated for its photo-actuation properties is polystyrene-*block*-poly(vinylmethylsiloxane)-*block*-polystyrene triblock copolymer [45]. In order to introduce light triggered groups into the matrix, the poly(vinylmethylsiloxane) block was covalently modified by attaching of the pendant azobenzene groups (Figure 13). This material exhibited two T_g (23°C and 100°C for azobenzene-modified silane block and polystyrene block, respectively). Such narrow difference between the T_g provides very narrow

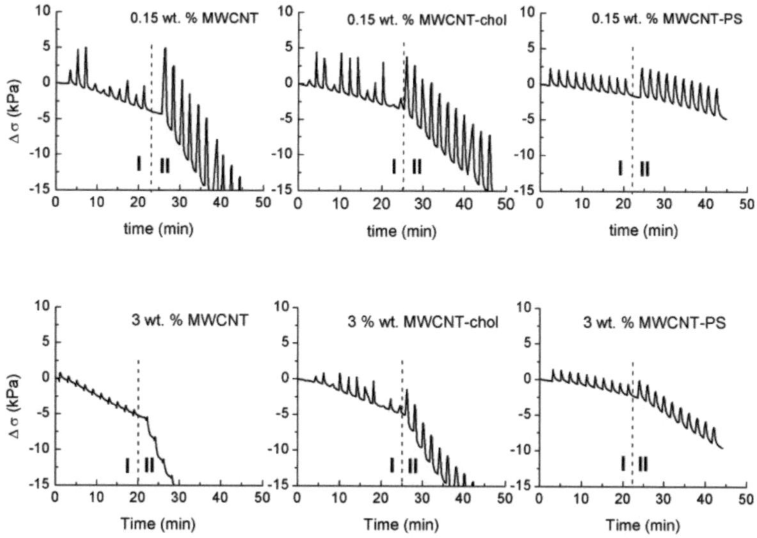

Figure 12. Changes in photo-actuation stress for composites containing either 0.15 wt. % or 3 wt. % of neat MWCNT or cholesteryl-modified multiwalled carbon nanotubes (MWCNT) (MWCNT-chol) or polystyrene-modified MWCNT (MWCNT-PS) during irradiation for 10 s (region I) and for 30 s (region II). Reprinted with permission from Ilcikova M, Mosnacek J, Mrlik M, Sedlacek T, Csomorova K, Czanikova K, et al. Influence of surface modification of carbon nanotubes on interactions with polystyrene-b-polyisoprene-b-polystyrene matrix and its photo-actuation properties. Polymers for Advanced Technologies. 2014;25(11):1293-300. Copyright © 2014 John Wiley & Sons, Ltd. [72].

Matrix / Filler	T_g (°C) soft phase / hard phase				E_a (kJ/mol) soft/hard phase
	0.5 Hz	**1 Hz**	**2.5 Hz**	**5 Hz**	
Pure SIS	-50.5 / 110.3	-48.7 / 111.1	-46.4 / 112.8	-45.0 / 116.2	174 / 475
SIS / MWCNT	-51.7 /102.8	-49.0 / 105.5	-47.8 / 109.9	-46.5 / 111.8	186 / 310
SIS/ MWCNT-chol	-47.8 / 110.3	-47.0 / 110.7	-45.0 / 115.2	-44.0 / 117.1	250 / 348
SIS/ MWCNT-PS	-50.4 / 111.8	-48.6 / 112.6	-46.4 / 114.3	-45.5 / 116.1	191 / 650
Pure MBM	-35.5 / NA	-33.7 / NA	-31.3 / NA	-29.5 / NA	184 / NA
MBM /MWCNT	-34.1 / NA	-33.3 / NA	-30.9 / NA	-25.9 / NA	196 / NA
MBM/ MWCNT-*g*-PBA	-34.0 / NA	-30.8 / NA	-27.3 / NA	-25.8 / NA	209 / NA
MBM/ MWCNT-*g*-PBA-*b*-PMMA	-30.8 / NA	-28.7 / NA	-27.1 / NA	-25.5 / NA	223 / NA

Table 1. Glass transition temperatures (T_g), obtained at various frequencies from DMA measurements and corresponding calculated activation energies of glass transitions of individual phases for pure polystyrene-*block*-polyisoprene-*block*-polystyrene (SIS) and pure poly(methyl methacrylate)- *block*-poly(butyl acrylate)-*block*-poly(methyl methacrylate) (MBM) matrices and their composites containing 1 wt. % of neat or modified MWCNTs. NA stays for Not Available, since T_g of PMMA phase was not observable.

application window for these materials, even though they provide very promising photo-actuation performance with reversible strain of 3.1% and tensile strength 25.7 kPa.

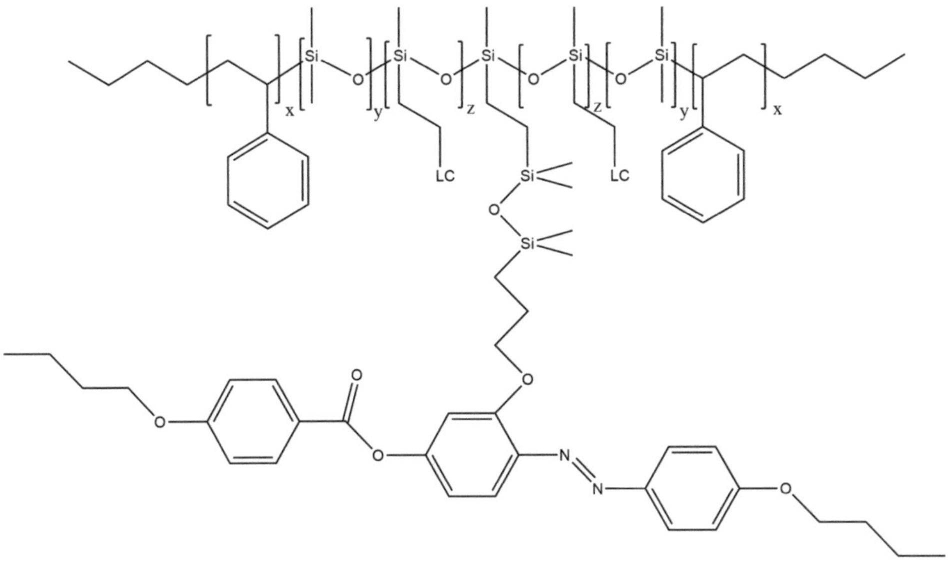

Figure 13. Schematic illustration of triblock copolymer TPEs containing pendant azobenzene groups [45].

3.5. Acrylic block copolymers

Acrylic-based TPEs are triblock copolymers consisting of one middle soft polyacrylate and two hard polymethacrylate blocks. These materials were just very recently investigated as photo-actuators with very promising results [48, 73]. Their main advantage compared with the styrene-based triblock TPEs is the higher UV stability, better mechanical properties at elevated temperatures, and variability of their properties depending on the choice of (meth)acrylate monomers structure.

The photo-actuation phenomenon has been investigated on poly(methyl methacrylate)-*block*-poly(butyl acrylate)-*block*-poly(methyl methacrylate) triblock copolymer (PMMA-*b*-PBA-*b*-PMMA) filled with MWCNTs as a light-triggered material. In addition to neat MWCNTs, also MWCNTs grafted with either poly(butyl acrylate) homopolymer (MWCNT-*g*-PBA) or poly(butyl acrylate)-*block*-poly(methyl methacrylate) diblock copolymer (MWCNT-*g*-PBA-*b*-PMMA) have been incorporated into the PMMA-*b*-PBA-*b*-PMMA matrix [48, 73]. The modification of MWCNTs was focused on the improvement of interactions with soft PBA phase, stiffening of the soft phase, and efficient heat transfer from MWCNTs to the soft PBA phase. Composites containing 1 wt. % of neat MWCNTs or MWCNT-*g*-PBA were prepared by solution mixing with polymer matrix, while 1 wt. % MWCNT-*g*-PBA-*b*-PMMA composite was prepared during *in situ* polymerization of PMMA-*b*-PBA-*b*-PMMA matrix (Figure 14) [48]. Therefore, the composites based on PMMA-*b*-PBA-*b*-PMMA containing 1 wt. % MWCNT-*g*-PBA-*b*-PMMA were prepared. It has been proved by electron microscopies that significantly better dispersity with more homogeneous distribution of the MWCNTs was obtained in the composite prepared *in situ* during polymerization. The modification of MWCNTs led to enhancement of both G′ and G′′ in wide range of temperatures (up to 260°C) compared to pure matrix, while the enhancement was most pronounced in the case of MWCNT-*g*-PBA-*b*-PMMA composite.

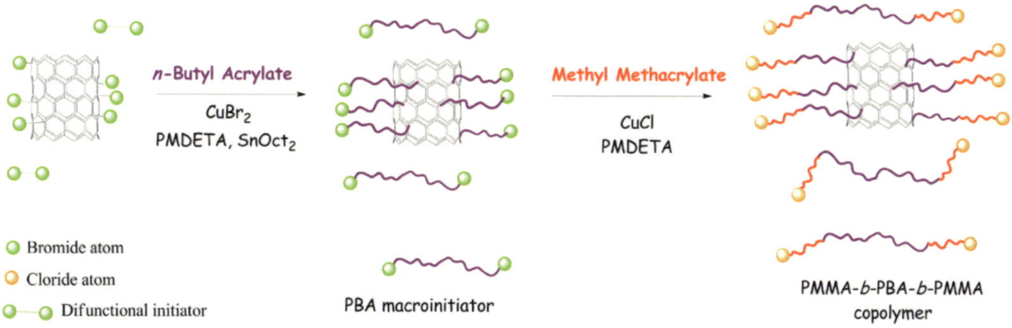

Figure 14. Schematic illustration of covalent modification of MWCNT surface with PBA-*b*-PMMA diblock copolymer *in situ* during synthesis of PMMA-*b*-PBA-*b*-PMMA matrix [48].

DMA analysis reveals that the incorporation of neat MWCNTs affected the T_g of PBA phase only negligibly (-33.3°C compared with -33.7°C for pure matrix). In the MWCNT-*g*-PBA composite the T_g increased to -30.8°C and the most pronounced shift was obtained in the case of MWCNT-*g*-PBA-*b*-PMMA composite (-28.7°C) [73]. The activation energies of glass transition of the PBA phase increased in the same order (see Table 1). Similarly, significantly improved photo-actuation ability was obtained in the case of MWCNT-*g*-PBA-*b*-PMMA composites (Figure 15). Under all investigated conditions the reversible contraction with the highest absolute values of photo-actuation was observed in this composite. It exhibited the actuation contraction of 280 μm and 400 μm at light power of 6.6 mW and 18.5 mW, respectively, after 30 s of irradiation. These values correspond to 3.2% and 4.5% change in sample length, respectively. Since both MWCNT-*g*-PBA and MWCNT-*g*-PBA-*b*-PMMA were found to interact preferentially with soft PBA phase, the big difference between the photo-actuation behavior of these two composites can be assigned to much better dispersity and more homogeneous distribution of the later one resulting in more effective heat transfer to the matrix.

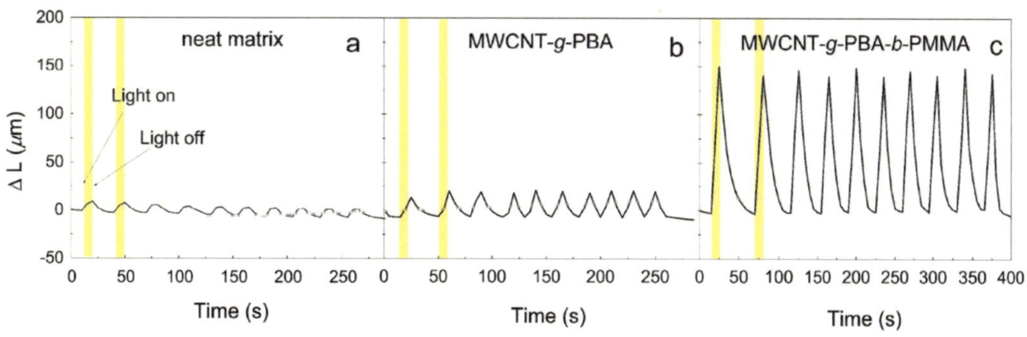

Figure 15. Comparison of actuation length change of neat matrix (a), composite containing 1 wt. % of MWCNT-g-PBA (b) and MWCNT-*g*-PBA-*b*-PMMA (c), pre-load: 0.05N, irradiation for 10 s, light power of 6.6 mW [73].

4. Methods for determination of photo-actuation behavior

4.1. Setups for measuring of photo-actuation performance

Several different methods have been developed to measure photo-actuation ability of TPE materials. A setup frequently used for the investigation of photo-actuation behavior is the dynamometer [11, 50]. In this case, the sample in the form of the stripe is fixed in the upper holder and in the bottom holder. On the bottom holder, the weight of various values is mounted. Thus, the certain pre-load is applied (Figure 16). The pre-loaded sample is then exposed to photo-stimulation and the change in the length is measured. The main advantage is the relatively easy construction of the setup; however, if the materials with different mechanical properties are subjected to the certain pre-load (depending on used weights), different pre-strain is achieved. Hence, the adjusting to certain pre-strain is rather impossible using various weights. Also, the accuracy of the length change measurement is not precise [11].

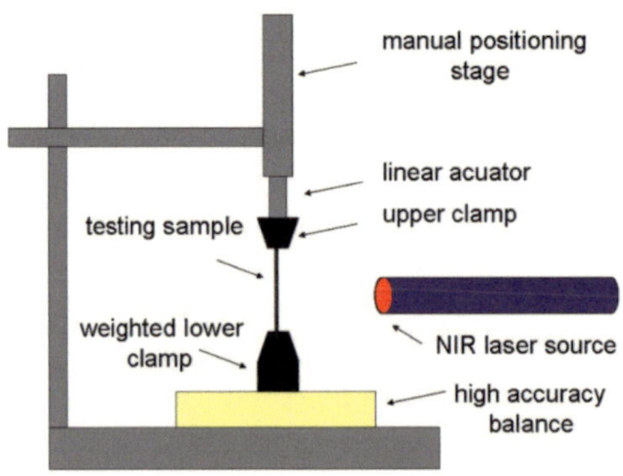

Figure 16. Dynamometer setup used for the investigation of photo-actuation behavior [11].

The same principle of applying certain pre-load has been used in the case of samples measured with the thermo-mechanical analyzer (TMA) [48, 73]. Compared to the previous setup, this device is very accurate and a certain value of pre-strain can be finely tuned. The TMA device is able to collect the data each 0.1 s depending on the settings and is able to record the change in the length automatically with very high precision usually in nanometers scale.

A very often utilized device for the investigation of photo-actuation behavior of TPE-based actuators in the form of the stripes is the universal tensile testing machine [12]. This machine allows to define a certain level of the pre-strain and thus a certain degree of alignment of polymer chains can be achieved. In the case of TPU samples, strain-induced crystallization also takes place and is fixed. When the material reaches equilibrium, it is able to respond on application of the light stimulus and to exhibit photo-actuation phenomenon [12].

The most precise device for the photo-actuation investigation is dynamic mechanical analysis (DMA) in iso-strain tensile mode (Figure 17) [50, 72]. The sample in the form of the strip is fixed into the clamps and the certain pre-strain is set. After stabilization of sample stress while keeping the same clamp distance, photo-stimulation is started. This device very precisely

http://dx.doi.org/10.5772/59647

calculates the position of the clamps and records the load resulting from the contraction or expansion of the sample. According to these values, the device automatically provides the certain level of the actuation displacement (in the range of μm) of the sample, as well as the actuation stress (in the range of kPa) as the most important values for the evaluation of the photo-actuation phenomenon.

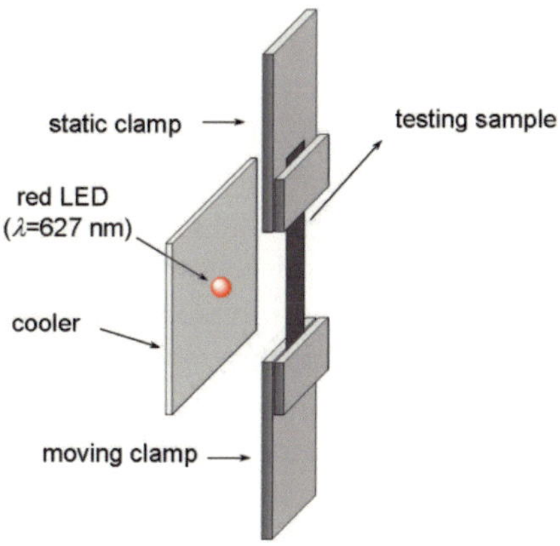

Figure 17. DMA setup for investigation of the photo-actuation performance [50].

Some photo-actuation studies were performed directly for the possible application of the materials in the development of new types of tactile displays. In these studies, the Braille-like elements were prepared instead of stripes and thus, also, different methods for evaluation of their photoactuation behavior have been developed. Atomic force microscopy (AFM) was one of the utilized devices to detect the movement of the Braille-like element (Figure 18). In this method, the AFM tip is placed on the top of the element and after exposition to the light the AFM records the tip movement up to several μm [46, 47]. This technique is very useful for the investigation of the photo-actuation kinetics; however, the actuation stress cannot be calculated and the actuation displacement investigation is limited by maximal possible displacement of AFM tip.

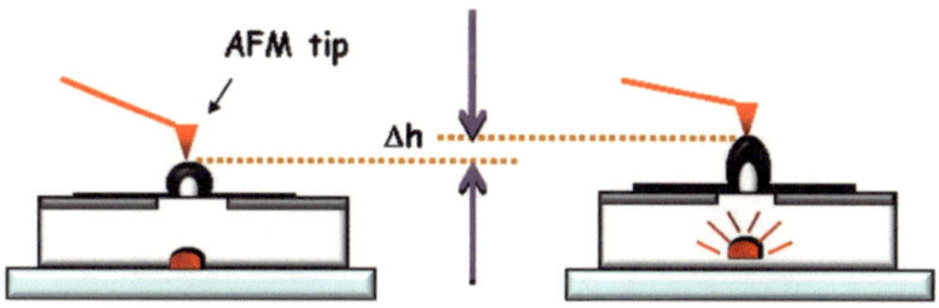

Figure 18. AFM setup for investigation of the photo-actuation performance of the samples in the form of Braille-like elements.

The second device utilized for the investigation of the Braille-like elements is scanning electron microscopy (SEM) [47, 49]. In this case, the Braille-like element is placed into the SEM evacuated chamber and then the samples are irradiated through the self-modified setup. The maximal change in the Braille-like element height is recorded. Again, this method does not allow the calculation of the actuation stress, but it can provide good information of the Braille-like element photo-actuation performance.

4.2. Preparation of samples for measuring of photo-actuation performance

In most of the set-ups used for the measurement of photo-actuation, the tested samples were in the form of strips. The polymer films for strips can be simply prepared by casting from the solution. However, very often, carbon base fillers are used as light-triggered materials that have to be incorporated into the TPE matrix. In order to provide the photo-actuator with promising behavior, the additive should be well dispersed and homogenously distributed in the TPE matrix. There were two methods reported for preparation of photo-actuating TPE composites in the literature. The first method is solution mixing, where the filler is dispersed in the proper solvent, while very often the ultrasonication is needed, and TPE matrix is afterwards added to the additive dispersion either in the form of solid or solution. When TPE is added as a solid, the mixture has to be mixed for a sufficiently long time to completely dissolve the matrix [47–51, 72]. Sonication and/or high shear mixing can be applied in order to disperse the filler well [46, 72].

The second method is based on the preparation of *in situ* composites during synthesis of polymer matrix in the presence of a filler [48, 73]. This method allows direct modification of filler surface during the polymerization process and can provide composites with much better dispersed filler. After polymerization, the polymer/filler mixture has to be precipitated to remove unreacted monomer and then re-dispersed in a solvent for casting process. In both methods, finally, the homogenous mixture is casted onto the Teflon dishes or molds of specified shape [50, 51] and the solvent is evaporated at elevated temperature and reduced pressure. Then the samples for photo-actuating investigation is cut to obtain the regular shape of stripes not exceeding dimensions of length 30 mm, width 10 mm, and thickness 1 mm. The thickness of the tested stripes was usually in the range of 0.3 mm–0.8 mm [47–51, 72].

In the case of Braille-like elements, thermoforming using a specific mold was used (Figure 19). Thermoforming allows not only preparation of the sample with required shape, but also obtaining certain pre-strain and orientation of the polymer chains needed for achieving the photo-actuation behavior of the material [46, 47].

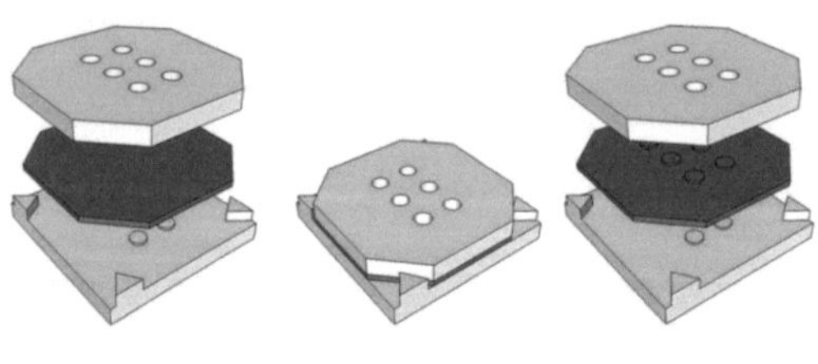

Figure 19. Thermoforming process of the Braille-like elements. Reprinted with permission from Ilcikova M, Mrlik M, Sedlacek T, Chorvat D, Krupa I, Slouf M, et al. Viscoelastic and photo-actuation studies of composites based on polystyrene-grafted carbon nanotubes and styrene-b-isoprene-b-styrene block copolymer. Polymer. 2014;55(1):211-8. Copyright © 2013 Elsevier Ltd. [46].

5. Applicability of photo-actuating TPE materials

The biocompatible grades of TPU open opportunity for utilization of photo-thermal activated actuators for *in vivo* medical applications. One of the proposed actuator was a photo-thermally expandable vascular stent for treatment of arterial stenosis. The polymer-based stents were designed to replace the frequently used metallic stents due to enhanced flexibility and possible drug elution [1]. Engineering aspects related to the application of photo-actuating thermoset TPU as materials for intravascular laser activated devices were reported. There were two type of devices designed. The interventional ischemic stroke devices and micro-grippers based on releasing embolic coils. The crucial parameter of suitable TPU stents was addressed to T_g that has to be low enough to actuate at the lowest laser power, but it should be high enough not to self-expand at body temperature. The suitable materials were determined as those actuated in the range of 65°C to 80°C. The crucial issue was determined as attaining these temperatures in the actuators in flow conditions and coupling of the light from the diffusing fiber through the blood into the actuator.

Later on, laser actuated intravascular thrombectomy device based on thermoplastic TPU has been designed [2, 3]. The prototype of self-expanding stent was composed of TPU crimped over a light diffuser attached to the end of the optical fiber connected to the IR laser diode [3]. The optical fiber had a dual function, first as a transport vehicle and second as a light energy delivery. The authors used the MM5520 phase separated TPU with T_gs of -25°C and 55°C. The material was first crimped above T_g, i.e., above 55°C, and cooled down. The primary shape was achieved after heating the material to 40°C–45 °C. The modulus turned from 800 MPa to 1.4 MPa during the shape change. The requirements for the moduli of the expandable polymer stents lie in the range from 800 MPa to 7 GPa in the glassy state. The applications such as neural stenting require low mechanical strength and the moduli ranges from 400 MPa in the glassy state. However, in this case, the low expansion force is generated [4].

The TPU was considered as a promising candidate for vascular stents, as the requirements for mechanical properties, deployment, and biocompatibility are met. Nevertheless, there are still some aspects to face, such as those related to the prevention of photo-thermally induced injury of arterial tissue.

Recently, the photo-thermally responsive TPEs have been utilized in the development of tactile displays for blind or visually impaired people [5]. The everyday activities of the population of the whole world can be greatly improved with the implementation of at least one Braille-like element in various devices. For wide industrial application, the cost of raw material and processing belong to decisive factors. From this point of view, the TPE poses advantage over still costly LCs or chemically cross-linked elastomers, which were recently studied for the same application as well.

Another potential application in plastic motors has emerged recently [9]. The photo-thermally activated localized contraction of scrolled strip caused rolling of the sample upon irradiation by UV and visible light. The investigated material was based on LCE with aligned structure.

Even though TPE actuators have not been tested in this pioneer publication, it can be assumed that TPEs with oriented structure can be suitable candidates for such application as well. Similarly, even though so far TPE photo-actuators have not been investigated for other applications, these materials that have responses in visible or near-infrared (NIR) wavelength region have various potential applications in telecommunication, thermal imaging, remote sensing, thermal photovoltaics, solar cells, sensors, etc.

6. Conclusions

The photo-actuating phenomenon is mostly influenced by the chemical structure of matrix, light absorbers, and fillers improving mechanical properties and thermal conductivity of the final material. Thus, photo-actuating materials with good performance are mostly composites consisting of elastomers and carbon-based fillers. TPEs having physically cross-linked structures possess several advantages compared to chemically cross-linked elastomers. The advantages of TPEs compared to chemically cross-linked elastomers are repeatable processability, tuning their mechanical properties by slight changes in composition and chemical structure, and price. Good photo-actuation response has been shown for TPE composites mainly based on thermoplastic urethanes, poly(ethylene-*co*-vinyl acetate), polystyrene-*block*-polyisoprene-*block*-polystyrene and poly(methyl methacrylate)-*block*-poly(butyl acrylate)-*block*-poly(methyl methacrylate) matrices. Carbon nanotubes and graphene are appropriate fillers providing good optical, mechanical, and thermal properties at quite low loading. Good dispersity and homogeneous distribution of the filler in the TPE matrix is crucial for achieving materials with good photo-actuation behavior. Therefore, surface modification of the fillers to improve compatibility with TPE matrix is encouraged. In block copolymers-based TPEs, preferential interactions of fillers with soft phase should be tailored to maximize the heat energy transfer to the soft phase, which is responsible for the actuation changes of the material. The photo-actuation phenomenon has been utilized in various smart applications. Poly(ethylene-*co*-vinyl acetate) and polystyrene-*block*-polyisoprene-*block*-polystyrene have been used in the development of tactile displays, and the biocompatible thermoplastic polyurethanes have been utilized for fabrication of vascular stents. The applicability of these materials is, however, much broader than published so far.

Acknowledgements

The authors thank the Centre of Excellence SAS for Functionalized Multiphase Materials (FUN-MAT), Grant agency VEGA 2/0112/13, Slovak Research and Development Agency APVV through Grant APVV-0109-10 and project SAS-MOST JRP 2014-9 "Synthesis of well-defined novel copolymers by use of living polymerization methods and advanced chromatography technique".

Author details

Markéta Ilčíková[1], Miroslav Mrlík[1] and Jaroslav Mosnáček[2*]

*Address all correspondence to: upolmosj@savba.sk

1 Center for Advanced Materials, Qatar University, Doha, Qatar

2 Polymer Institute, Slovak Academy of Sciences, Bratislava, Slovakia

References

[1] Ahir SV, Huang YY, Terentjev EM. Polymers with aligned carbon nanotubes: Active composite materials. Polymer. 2008;49(18):3841-54, DOI:10.1016/j.polymer. 2008.05.005.

[2] Lendlein A, Jiang H, Junger O, Langer R. Light-induced shape-memory polymers. Nature. 2005;434(7035):879-82.

[3] Lendlein A, Kelch S. Shape-memory polymers. Angewandte Chemie International Edition. 2002;41(12):2034-57.

[4] Lau GK, Goosen JFL, Keulen Fv, Duc TC, Sarro PM. Powerful polymeric thermal microactuator with embedded silicon microstructure. Appl Phys Lett. 2007;90:214103.

[5] Shankar R, Ghosh TK, Spontak RJ. Electromechanical Response of Nanostructured Polymer Systems with no Mechanical Pre-Strain. Macromol Rapid Commun. 2007;28(10):1142-7.

[6] Shankar Ravi, Ghosh Tushar K., J. SR. Dielectric elastomers as next-generation polymeric actuators. Soft Matter. 2007;3:1116-29.

[7] Lu S, Panchapakesan B. Photomechanical responses of carbon nanotube/polymer actuators. Nanotechnology. 2007;18:305502.

[8] Lu S, Ahir SV, Terentjev EM, Panchapakesan B. Alignment dependent mechanical responses of carbon nanotubes to light. Appl Phys Lett. 2007;91:103106.

[9] Garcia-Amorós J, Finkelmann H, Velasco D. Increasing the isomerisation kinetics of azo dyes by Chemical bonding to liquid-crystalline polymers. Chemistry – A European Journal. 2011;17(23):6518-23.

[10] Li F, Qi L, Yang J, Xu M, Luo X, Ma D. Polyurethane/conducting carbon black composites: Structure, electric conductivity, strain recovery behavior, and their relationships. J Appl Polym Sci. 2000;75(1):68-77.

[11] Loomis J, King B, Burkhead T, Xu P, Bessler N, Terentjev E, et al. Graphene-nanoplatelet-based photomechanical actuators. Nanotechnology. 2012;23(4):045501, DOI: 10.1088/0957-4484/23/4/045501.

[12] Koerner H, Price G, Pearce NA, Alexander M, Vaia RA. Remotely actuated polymer nanocomposites - stress-recovery of carbon-nanotube-filled thermoplastic elastomers. Nature Materials. 2004;3(2):115-20, DOI:10.1038/nmat1059.

[13] Habault D, Zhang H, Zhao Y. Light-triggered self-healing and shape-memory polymers. Chemical Society Reviews. 2013;42(17):7244-56, DOI:10.1039/c3cs35489j.

[14] Tong ZZ, Xue JQ, Wang RY, Huang J, Xu JT, Fan ZQ. Hierarchical self-assembly, photo-responsive phase behavior and variable tensile property of azobenzene-containing ABA triblock copolymers. Rsc Advances. 2015;5(6):4030-40, DOI:10.1039/c4ra12844c.

[15] Zhao JQ, Liu YY, Yu YL. Dual-responsive inverse opal films based on a crosslinked liquid crystal polymer containing azobenzene. Journal of Materials Chemistry C. 2014;2(48):10262-7, DOI:10.1039/c4tc01825g.

[16] Finkelmann H, Nishikawa E, Pereira GG, Warner M. A new opto-mechanical effect in solids. Physical Review Letters. 2001;87(1):015501, DOI:10.1103/PhysRevLett.87.015501.

[17] Ziolkowski B, Florea L, Theobald J, Benito-Lopez F, Diamond D. Self-protonating spiropyran-co-NIPAM-co-acrylic acid hydrogel photoactuators. Soft Matter. 2013;9(36):8754-60, DOI:10.1039/c3sm51386f.

[18] Priimagi A, Barrett CJ, Shishido A. Recent twists in photoactuation and photoalignment control. Journal of Materials Chemistry C. 2014;2(35):7155-62, DOI:10.1039/c4tc01236d.

[19] Yu HF, Ikeda T. Photocontrollable liquid-crystalline actuators. Advanced Materials. 2011;23(19):2149-80, DOI:10.1002/adma.201100131.

[20] White TJ, McConney ME, Bunning TJ. Dynamic color in stimuli-responsive cholesteric liquid crystals. Journal of Materials Chemistry. 2010;20(44):9832-47, DOI:10.1039/c0jm00843e.

[21] Coles H, Morris S. Liquid-crystal lasers. Nature Photonics. 2010;4(10):676-85, DOI: 10.1038/nphoton.2010.184.

[22] Ikeda T, Nakano M, Yu YL, Tsutsumi O, Kanazawa A. Anisotropic bending and unbending behavior of azobenzene liquid-crystalline gels by light exposure. Advanced Materials. 2003;15(3):201, DOI:10.1002/adma.200390045.

[23] Finkelmann H, Kock HJ, Rehage G. Investigations on liquid-crystalline polysiloxanes.3. liquid-crystalline elastomers - A new type of liquid-crystalline material. Mak-

romolekulare Chemie-Rapid Communications. 1981;2(4):317-22, DOI:nema to DOI na WOS.

[24] Warner M, Terentjev EM. Liquid Crystal Elastomers. Clarendon Press, Oxford. 2003, DOI:10.1002/pat.1294.

[25] Yu YL, Nakano M, Ikeda T. Directed bending of a polymer film by light - Miniaturizing a simple photomechanical system could expand its range of applications. Nature. 2003;425(6954):145, DOI:10.1038/425145a.

[26] Camacho-Lopez M, Finkelmann H, Palffy-Muhoray P, Shelley M. Fast liquid-crystal elastomer swims into the dark. Nature Materials. 2004;3(5):307-10, DOI:10.1038/nmat1118.

[27] Koerner H, White TJ, Tabiryan NV, Bunning TJ, Vaia RA. Photogenerating work from polymers. Materials Today. 2008;11(7-8):34-42, DOI:10.1016/s1369-7021(08)70147-0.

[28] Yamada M, Kondo M, Mamiya JI, Yu YL, Kinoshita M, Barrett CJ, et al. Photomobile polymer materials: Towards light-driven plastic motors. Angewandte Chemie-International Edition. 2008;47(27):4986-8, DOI:10.1002/anie.200800760.

[29] Yin RY, Xu WX, Kondo M, Yen CC, Mamiya J, Ikeda T, et al. Can sunlight drive the photoinduced bending of polymer films? Journal of Materials Chemistry. 2009;19(20):3141-3, DOI:10.1039/b904973h.

[30] Priimagi A, Shimamura A, Kondo M, Hiraoka T, Kubo S, Mamiya JI, et al. Location of the azobenzene moieties within the cross-linked liquid-crystalline polymers can dictate the direction of photoinduced bending. ACS Macro Letters. 2012;1(1):96-9, DOI:10.1021/mz200056w.

[31] van Oosten CL, Bastiaansen CWM, Broer DJ. Printed artificial cilia from liquid-crystal network actuators modularly driven by light. Nature Materials. 2009;8(8):677-82, DOI:10.1038/nmat2487.

[32] Eisenbach CD. Isomerization of aromatic azo chromophores in poly(ethyl acrylate) networks and photomechanical effect. Polymer. 1980;21(10):1175-9, DOI:10.1016/0032-3861(80)90083-x.

[33] Garcia-Amoros J, Martinez M, Finkelmann H, Velasco D. Photoactuation and thermal isomerisation mechanism of cyanoazobenzene-based liquid crystal elastomers. Physical Chemistry Chemical Physics. 2014;16(18):8448-54, DOI:10.1039/c4cp00446a.

[34] Li CS, Liu Y, Lo CW, Jiang HR. Reversible white-light actuation of carbon nanotube incorporated liquid crystalline elastomer nanocomposites. Soft Matter. 2011;7(16):7511-6, DOI:10.1039/c1sm05776f.

[35] Li ME, Lv S, Zhou JX. Photo-thermo-mechanically actuated bending and snapping kinetics of liquid crystal elastomer cantilever. Smart Mater Struct. 2014;23(12), DOI: 10.1088/0964-1726/23/12/125012.

[36] Hamon MA, Itkis ME, Niyogi S, Alvaraez T, Kuper C, Menon M, et al. Effect of rehybridization on the electronic structure of single-walled carbon nanotubes. J Am Chem Soc. 2001;123(45):11292-3, DOI:Doi 10.1021/Ja0109702.

[37] Yang LQ, Setyowati K, Li A, Gong SQ, Chen J. Reversible infrared actuation of carbon nanotube-liquid crystalline elastomer nanocomposites. Advanced Materials. 2008;20(12):2271, DOI:10.1002/adma.200702953.

[38] Ji Y, Huang YY, Rungsawang R, Terentjev EM. Dispersion and Alignment of carbon nanotubes in liquid crystalline polymers and elastomers. Advanced Materials. 2010;22(31):3436, DOI:10.1002/adma.200904103.

[39] Ahir SV, Squires AM, Tajbakhsh AR, Terentjev EM. Infrared actuation in aligned polymer-nanotube composites. Physical Review B. 2006;73(8):085420, DOI:10.1103/PhysRevB.73.085420.

[40] Wang D, Ji WX, Li ZC, Chen LW. A biomimetic "polysoap" for single-walled carbon nanotube dispersion. J Am Chem Soc. 2006;128(20):6556-7, DOI:Doi 10.1021/Ja060907i.

[41] Ansari S, Varghese JM, Dayas KR. Polydimethylsiloxane-cristobalite composite adhesive system for aerospace applications. Polymers for Advanced Technologies. 2009;20(5):459-65, DOI:10.1002/pat.1294.

[42] Loomis J, King B, Panchapakesan B. Layer dependent mechanical responses of graphene composites to near-infrared light. Applied Physics Letters. 2012;100(7):073108, DOI:10.1063/1.3685479.

[43] Ahir SV, Terentjev EM. Photomechanical actuation in polymer-nanotube composites. Nature Materials. 2005;4(6):491-5, DOI:10.1038/nmat1391.

[44] Ahir SV, Terentjev EM. Fast relaxation of carbon nanotubes in polymer composite actuators. Physical Review Letters. 2006;96(13):133902, DOI:10.1103/PhysRevLett.96.133902.

[45] Petr M, Katzman BA, DiNatale W, Hammond PT. Synthesis of a new, low-t-g siloxane thermoplastic elastomer with a functionalizable backbone and its use as a rapid, room temperature photoactuator. Macromolecules. 2013;46(7):2823-32, DOI:10.1021/ma400031z.

[46] Ilcikova M, Mrlik M, Sedlacek T, Chorvat D, Krupa I, Slouf M, et al. Viscoelastic and photo-actuation studies of composites based on polystyrene-grafted carbon nanotubes and styrene-b-isoprene-b-styrene block copolymer. Polymer. 2014;55(1):211-8, DOI:10.1016/j.polymer.2013.11.031.

[47] Czanikova K, Ilcikova M, Krupa I, Micusik M, Kasak P, Pavlova E, et al. Elastomeric photo-actuators and their investigation by confocal laser scanning microscopy. Smart Mater Struct. 2013;22(10):104001, DOI:10.1088/0964-1726/22/10/104001.

[48] Ilcikova M, Mrlik M, Sedlacek T, Slouf M, Zhigunov A, Koynov K, et al. Synthesis of Photoactuating Acrylic Thermoplastic Elastomers Containing Diblock Copolymer-Grafted Carbon Nanotubes. Acs Macro Letters. 2014;3(10):999-1003, DOI:10.1021/mz500444m.

[49] Czanikova K, Krupa I, Račko D, Šmatko V, Campo EM, Pavlova E, et al. In situ electron microscopy of Braille microsystems: photo-actuation of ethylene vinyl acetate/carbon nanotube composites. Materials Research Express. 2015;2:025601, DOI: 10.1088/2053-1591/2/2/025601.

[50] Czanikova K, Torras N, Esteve J, Krupa I, Kasak P, Pavlova E, et al. Nanocomposite photoactuators based on an ethylene vinyl acetate copolymer filled with carbon nanotubes. Sensors and Actuators B-Chemical. 2013;186:701-10, DOI:10.1016/j.snb.2013.06.054.

[51] Czanikova K, Krupa I, Ilcikova M, Kasak P, Chorvat D, Valentin M, et al. Photo-actuating materials based on elastomers and modified carbon nanotubes. Journal of Nanophotonics. 2012;6:063522, DOI:10.1117/1.jnp.6.063522.

[52] Viola EA, Levitsky IA, Euler WB. Kinetics of Photoactuation in Single Wall Carbon Nanotubeâ^'Nafion Bilayer Composite. The Journal of Physical Chemistry C. 2010;114(47):20258-66, DOI:10.1021/jp106653c.

[53] Levitsky IA, Kanelos PT, Woodbury DS, Euler WB. Photoactuation from a carbon nanotube-nafion bilayer composite. Journal of Physical Chemistry B. 2006;110(19): 9421-5, DOI:10.1021/jp0606154.

[54] Seneker SD, Born L, Schmelzer HG, Eisenbach CD, Fischer K. Diisocyanato Dicyclohexylmethane - Structure Property Relationships of Its Geometrical-Isomers in Polyurethane Elastomers. Colloid and Polymer Science. 1992;270(6):543-8, DOI:10.1007/bf00658285.

[55] Rohr J, Koenig K, Koepnick H, Seeman KH. Polyester. Ullmans Encyklopaedie der tehnischen Chemie, Auflage, Verlag Chemie, Weinheim. 1980.

[56] Smith CP, Reisch JW, Oconnor JM. Thermoplastic polyurethane elastomers made from high-molecular-weight poly-l(r) polyols. Journal of Elastomers and Plastics. 1992;24(4):306-22, DOI:10.1177/009524439202400404.

[57] Holden G, Kricheldorf HR, Quirk RP. Thermoplastic polyurethane elastomers. In: Holden G, Kricheldorf HR, Quirk RP, editors. Thermoplastic Elastomers. 3 edition ed: Hanser Gardner Pubns; 2004. p. 15-45.

[58] Liang JJ, Xu YF, Huang Y, Zhang L, Wang Y, Ma YF, et al. Infrared-Triggered Actuators from Graphene-Based Nanocomposites. Journal of Physical Chemistry C. 2009;113(22):9921-7, DOI:10.1021/jp901284d.

[59] Park JH, Dao TD, Lee HI, Jeong HM, Kim BK. Properties of Graphene/Shape Memory Thermoplastic Polyurethane Composites Actuating by Various Methods. Materials. 2014;7(3):1520-38, DOI:10.3390/ma7031520.

[60] Feng YY, Qin MM, Guo HQ, Yoshino K, Feng W. Infrared-actuated recovery of polyurethane filled by reduced graphene oxide/carbon nanotube hybrids with high energy density. Acs Applied Materials & Interfaces. 2013;5(21):10882-8, DOI:10.1021/am403071k.

[61] Oh SM, Oh KM, Dao TD, Lee HI, Jeong HM, Kim BK. The modification of graphene with alcohols and its use in shape memory polyurethane composites. Polymer International. 2013;62(1):54-63, DOI:10.1002/pi.4366.

[62] Choi JT, Dao TD, Oh KM, Lee HI, Jeong HM, Kim BK. Shape memory polyurethane nanocomposites with functionalized graphene. Smart Mater Struct. 2012;21(7): 075017, DOI:10.1088/0964-1726/21/7/075017.

[63] Sahoo NG, Rana S, Cho JW, Li L, Chan SH. Polymer nanocomposites based on functionalized carbon nanotubes. Progress in Polymer Science. 2010;35(7):837-67, DOI: 10.1016/j.progpolymsci.2010.03.002.

[64] Balandin AA. Thermal properties of graphene and nanostructured carbon materials. Nature Materials. 2011;10(8):569-81, DOI:10.1038/nmat3064.

[65] Dreyer DR, Park S, Bielawski CW, Ruoff RS. The chemistry of graphene oxide. Chemical Society Reviews. 2010;39(1):228-40, DOI:10.1039/b917103g.

[66] Huang XY, Zhi CY, Jiang PK. Toward Effective Synergetic Effects from Graphene Nanoplatelets and Carbon Nanotubes on Thermal Conductivity of Ultrahigh Volume Fraction Nanocarbon Epoxy Composites. Journal of Physical Chemistry C. 2012;116(44):23812-20, DOI:10.1021/jp308556r.

[67] Aravind SSJ, Ramaprabhu S. Graphene-multiwalled carbon nanotube-based nanofluids for improved heat dissipation. Rsc Advances. 2013;3(13):4199-206, DOI: 10.1039/c3ra22653k.

[68] Chu K, Li WS, Jia CC, Tang FL. Thermal conductivity of composites with hybrid carbon nanotubes and graphene nanoplatelets. Applied Physics Letters. 2012;101(21): 211903, DOI:10.1063/1.4767899.

[69] Im H, Kim J. Thermal conductivity of a graphene oxide-carbon nanotube hybrid/epoxy composite. Carbon. 2012;50(15):5429-40, DOI:10.1016/j.carbon.2012.07.029.

[70] Shin MK, Lee B, Kim SH, Lee JA, Spinks GM, Gambhir S, et al. Synergistic toughening of composite fibres by self-alignment of reduced graphene oxide and carbon nanotubes. Nature Communications. 2012;3:650, DOI:10.1038/ncomms1661.

[71] Ansari S, Neelanchery MM, Ushus D. Graphene/Poly(styrene-b-isoprene-b-styrene) Nanocomposite Optical Actuators. Journal of Applied Polymer Science. 2013;130(6): 3902-8, DOI:10.1002/app.39666.

[72] Ilcikova M, Mosnacek J, Mrlik M, Sedlacek T, Csomorova K, Czanikova K, et al. Influence of surface modification of carbon nanotubes on interactions with polystyrene-b-polyisoprene-b-polystyrene matrix and its photo-actuation properties. Polymers for Advanced Technologies. 2014;25(11):1293-300, DOI:10.1002/pat.3324.

[73] Ilčíková M, Mrlík M, Sedláček T, Danko M, Doroshenko M, Koynov K., et al. Tailoring of viscoelastic performance of tri-block copolymer through surface modification of carbon nanotubes and impact on photoactuating properties. Polymer. 2015; DOI: 10.1016/j.polymer.2015.03.060.

Chapter 7

Thermoplastic Resins used in Dentistry

Lavinia Ardelean, Cristina Maria Bortun,
Angela Codruta Podariu and Laura Cristina Rusu

Additional information is available at the end of the chapter

http://dx.doi.org/10.5772/60931

Abstract

Thermoplastic materials such as polyamides (nylon), acetal resins, epoxy resins, polystyrene, polycarbonate resins, polyurethane and acrylic thermoplastic resins were introduced in dentistry as an alternative to classic resins, which have major disadvantages such as the toxicity of the residual monomer, awkward wrapping system and difficult processing.

Indications for thermoplastic resins include partial dentures, preformed clasps, partial denture frameworks, temporary or provisional crowns and bridges, full dentures, orthodontic appliances, anti-snoring devices, different types of mouth guards and splints. Some flexible myofunctional therapy devices, used for orthodontic purposes, may also be made of thermoplastic silicone polycarbonate-urethane.

The main characteristics of thermoplastic resins used in dentistry are as follows: they are monomer-free and consequently nontoxic and nonallergenic, they are injected by using special devices, they are biocompatible, they have enhanced esthetics and they are comfortable to wear.

Keywords: Thermoplastic resins, injection devices, metal-free removable partial dentures

1. Introduction

Continuous development and progress of polymer's industry with applications in general and dental medicine was of great importance for the health domain. Using various types of resins for restorations in the oral cavity is beneficial from childhood till geriatric period [1-4].

Thermopolymerizable acrylic resins were first used in dental technique in 1936, this being a great step forward. Acrylic resins are also known as polymethylmethacrylate or PMMA. These synthetically obtained materials can be modeled, packed or injected into molds during the plastic phase and become solid after chemical polymerization [5, 6]. Thermopolymerisable acrylic resins have many disadvantages as increased porosity, high water retention, volume variations and irritating effect due to the residual monomer, awkward wrapping system and difficult processing. Because of these, once polymers developed, alternative materials such as polyamides (nylon), acetal resins, epoxy resins, polystyrene, polycarbonate resins etc. [7-9] came on the market.

The main characteristics of the thermoplastic resins used in dentistry are as follows: they are monomer-free and consequently nontoxic and nonallergenic, they are injected by using special devices, they are biocompatible, they have enhanced esthetics and they are comfortable to wear [10].

2. Types of thermoplastic resins used in dentistry

The classification of resins according to DIN EN ISO 1567 is presented in Table 1:

Type	Class (manufacturing)	Group (presentation form)
Type 1	Thermopolymerizable resins (>65°C)	Group 1: bicomponent powder and liquid Group 2: monocomponent
Type 2	Autopolymerizable resins (<65°C)	Group 1: bicomponent powder and liquid Group 2: bicomponent powder and casting liquid
Type 3	Thermoplastic resins	Monocomponent system grains in cartridges
Type 4	Light-cured resins	Monocomponent system
Type 5	Microwave cured resins	Bicomponent system

Table 1. The classification of resins according to DIN EN ISO 1567

Among the technologies for manufacturing removable complete and partial dentures we distinguish: heat-curing, self-curing, injection, light-curing, casting and microwave use [11].

Thermoplastic resins may be classified by their composition, as acetal resins, polycarbonate resins (belonging to the group of polyester resins), acrylic resins and polyamides (nylons).

The use of thermoplastic resins in dental medicine is continuously growing. The material is thermally plasticized and no chemical reaction takes place. The injection of plasticized resins into a mold represents a new technology in manufacturing complete and removable partial dentures [12].

At present, due to successive alterations in the chemical composition, thermoplastic materials are suitable for manufacturing removable partial dentures with no metallic components, resulting in the so-called "metal-free removable partial dentures" [13].

Indications for thermoplastic resins include removable partial dentures, preformed clasps [14], partial denture frameworks, temporary or provisional crowns and bridges, complete dentures, orthodontic appliances, anti-snoring devices, different types of mouth guards and splints. Some flexible myofunctional therapy devices, used for orthodontic purposes, may also be made of thermoplastic silicone polycarbonate-urethane.

2.1. Thermoplastic acetal

Thermoplastic acetal is a poly(oxy-methylene)-based material, which as a homopolymer has good short-term mechanical properties but as a copolymer has better long-term stability [15]. Due to its resistance to wear and fracture, combined with a certain amount of flexibility, acetal resin is an ideal material for preformed clasps for partial dentures, single-pressed unilateral partial dentures, partial denture frameworks (Figure 1), provisional bridges, occlusal splints and implant abutments, artificial teeth for removable dentures and orthodontic appliances [16].

Because of their resistance to occlusal wear, acetal resins are also well suited for maintaining vertical dimension during provisional restorative therapy. Acetal is not translucent and does not match the esthetic appearance of thermoplastic acrylic and polycarbonate [17].

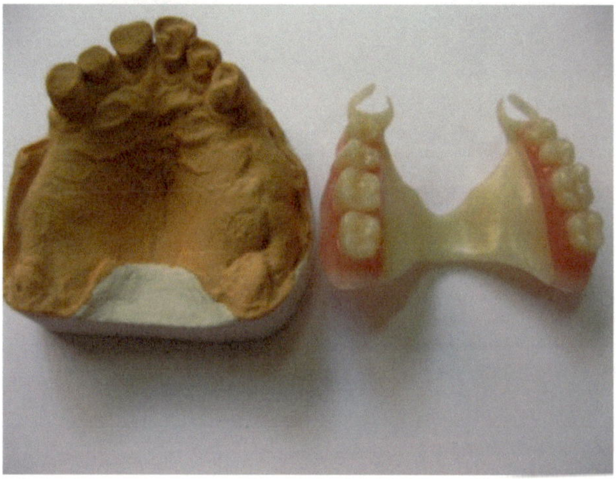

Figure 1. Removable partial denture with acetal frame and clasps

2.2. Thermoplastic polyamide (nylon)

Thermoplastic polyamide (nylon) is a resin derived from diamine and dibasic acid monomers. Versatililty is one of its characteristics and makes it suitable for various applications. Nylon exhibits high flexibility, physical strength, heat and chemical resistance. It can be easily modified to increase stiffness and wear resistance. Because of its excellent balance of strength, ductility and heat resistance, nylon is an outstanding candidate for metal replacement applications [18]. Nylon is mainly used for tissue supported removable dentures. Its stiffness makes it unsuitable for usage as occlusal rests or denture elements that need to be rigid [7, 13]. Because it is flexible, it cannot maintain vertical dimension when used in direct occlusal forces. Adjustment and polishing is difficult but provides excellent esthetics due to its semitranslucency [19, 20].

Nylon is specially indicated for patients allergic to methyl metacrylate, being monomer-free, lightweight and impervious to oral fluids [21]. Some may also be combined with a metal framework (Figure 2).

Figure 2. Removable partial denture of polyamide combined with metal

Comparative properties of thermoplastic acetal and polyamide, the two types of resins suitable for manufacturing removable partial dentures, are shown in Table 2.

Resin type	Main substance	Resistance	Durity	Flexibility	Esthetics	Biocompatibility
Acetalic resin	polioximetylen	very good	very high	medium	good	very good
Polyamidic resin	diamine	good	high	medium or very high	very good	very good

Table 2. Comparative aspects of acetalic and polyamidic thermoplastic resins

2.3. Thermoplastic polyester

Thermoplastic polyester resins are also used in dentistry. They melt between 230°C and 290°C and the technology implies casting into molds. Polycarbonate resins are particularly polyester materials. They have good fracture strength and flexibility, but the wear resistance is lower than acetal resins. Polycarbonates have a natural translucency and finishes very well, which recommends them for temporary restorations, but they are not suitable for partial denture frameworks [22].

2.4. Thermoplastic acrylate

Thermoplastic acrylate consists of fully polymerized acrylate, its base component being methyl-metacrylate, the special blend of polymers giving it the highest impact rating of any acrylic. This material was developed for manufacturing complete dentures. It is not elastic, but its flexibility makes it practically unbreakable. The material has long-term stability, its surface structure being dense and smooth. Due to the absence of residual monomer its biocompatibility is very good. The denture has very good long-term adaptability because water retention is limited. You can bounce such denture off the floor without cracking the base [7, 21, 23].

2.5. Presentation form and injection

Thermoplastic materials can be polymerized or prepolymerized and they can be found in granular form, with low molecular weight, already wrapped in cartridges that eliminate dosage errors (Figure 3).

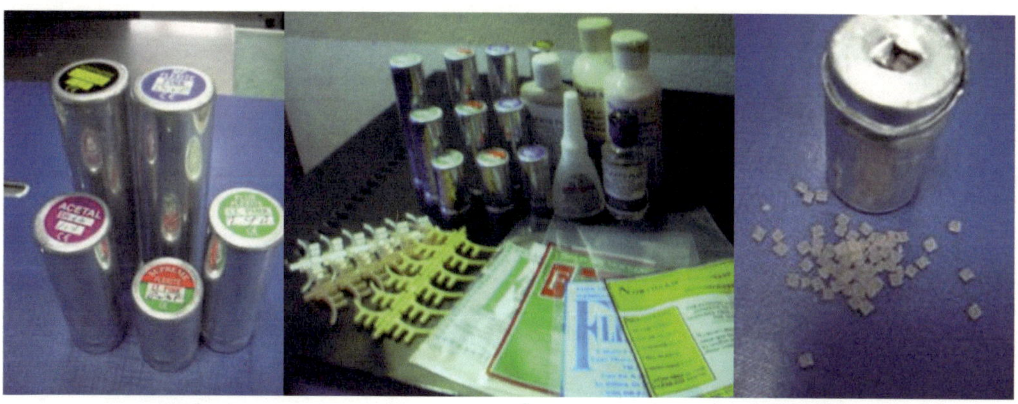

a. b. c.

Figure 3. (a, b) Cartridges of different thermoplastic resins (c) The granular aspect of the material

They exhibit a high rigidity despite their low molecular weight. Their plasticizing temperature is low (200°C-250°C). Thermal plasticization takes place in special devices afterward the material is injected under pressure into a mold, without any chemical reactions. After heating, the metallic cartridges containing thermoplastic grains are set in place into the injecting unit and the plasticized resin is forced into the mold at a pressure of 6-8 bars. Pressure, temperature and injecting time are automatically controlled by the injecting unit. Dentures obtained using this technology have excellent esthetics and good compatibility [7, 12, 13, 22].

Injecting thermoplastic resins into molds is not a common technology in dental laboratories because the need of expensive equipment and this could be a disadvantage.

The special injection devices we use are Polyapress (Bredent) and R-3C (Flexite) injectors (Figure 4).

a. b.

Figure 4. (a) The Polyapress injection-molding device (Bredent) (b) The R-3C injector (Flexite)

3. Prosthetic devices made of thermoplastic resins

The main characteristics of thermoplastic resins used in dentistry are as follows: they are monomer-free and consequently nontoxic and nonallergenic, they are injected by using special devices, they are biocompatible, they have enhanced esthetics and they are comfortable to wear.

Our experience with thermoplastic resins for dental use involves solving several different cases of partial edentations, with removable partial dentures without metallic framework, or combining the metallic framework with thermoplastic resin saddles, using different thermoplastic resins, selected according to their indications and manufacturing technology (Figure 5).

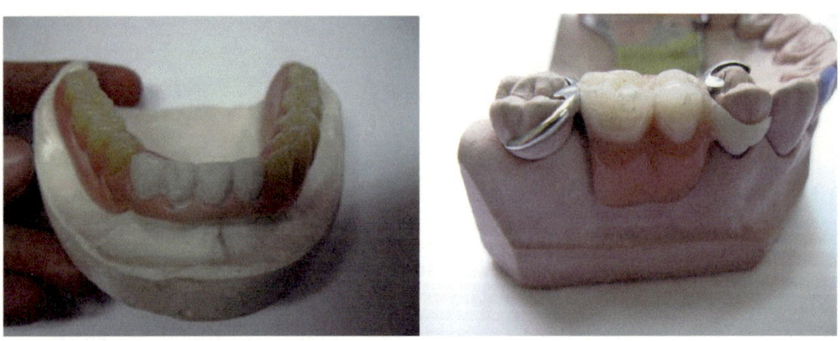

Figure 5. Different combinations between thermoplastic resins, with or without metal

The types of thermoplastic resins we used for manufacturing different types of removable partial dentures are acetal resins and polyamides of different flexibilities.

http://dx.doi.org/10.5772/60931

3.1. Removable partial dentures with acetal framework

The acetal resin has optimal physical and chemical properties and it is indicated in making frameworks and clasps for removable partial dentures, being available in tooth color and pink [12]. Experimentally, in some cases, we combined an acetal resin frames with classic acrylic resins for the saddles (Figure 4). However, the resistance values for the acetal resin framework do not reach those of a metal one [16], consequently the main connector, the clasps and the spurs need to be oversized [12]. Injection was carried out using the R-3 C digital control device that has five preset programs, as well as programs that can be individually set by the user.

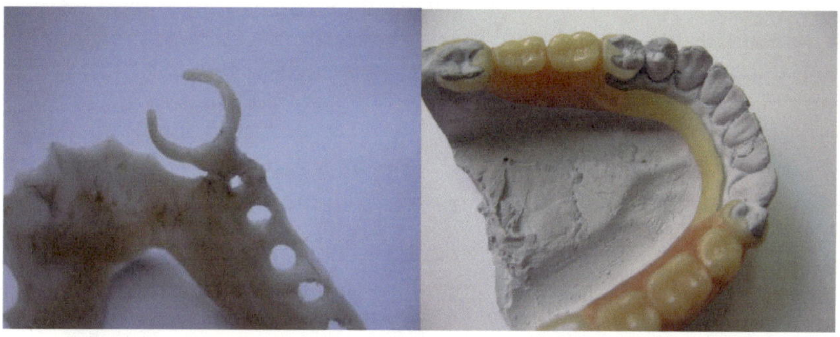

Figure 6. Acetal framework and clasp and removable partial dentures with acetal framework and clasps

The maintenance, support and stabilizing systems used are metal-free ones.

The significant aspects of the technical steps in the technology of removable partial dentures made of thermoplastic materials are described.

The master model is poured of class IV hard plaster, using a vibrating table (Figure 7).

Figure 7. Casting the master model

In order to assess its retentiveness and to determine the place where the active arms of the clasp are placed a parallelograph analysis is made (Figure 8). The abutment teeth are selected and the position of the cast is chosen and recorded so that a favorable path of insertion is obtained. To record the position of the cast tripod marks are used. The contour heights on the abutment teeth and the retentive muco-osseous tissues are marked. The abutments undercuts are measured and the engagement of the terminal third of the retentive arms of the clasps is established.

After the parallelograph analysis is carried out, the future frame design is drawn, including all extensions of saddles, major connector, retentive and bracing arms of the clasps, occlusal rests and minor connectors of Ackers circumferential clasps on abutment teeth. The design starts with the saddles, following the main connector, the retentive and opposing clasp arms, the spurs and the secondary connectors of the Ackers circular clasps [12, 13].

After designing the framework, the master model is prepared for duplication, including foliation and deretentivisation (Figure 9). At the beginning, blue wax plates are used as spacers in regions where the framework has to be spaced from the gingival tissue. [12].

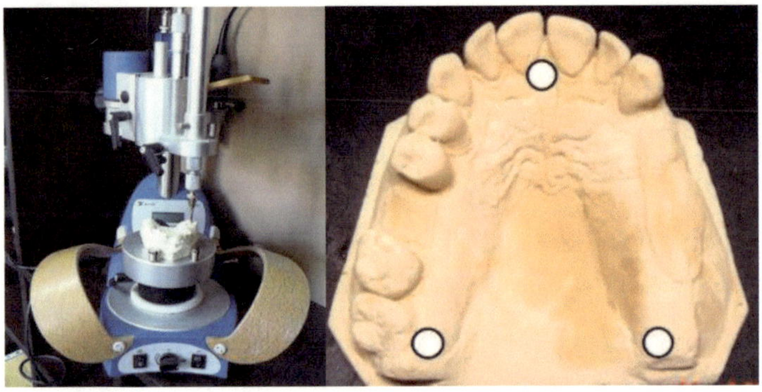

Figure 8. Parallelograph analysis

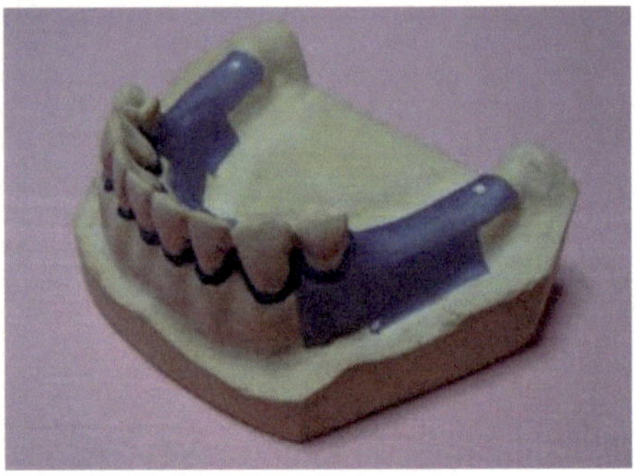

Figure 9. Deretentivisation of the model

http://dx.doi.org/10.5772/60931

Block-out wax is applied between teeth cervices and gingival margin of the drawing representing the clasps arms. The block-out wax meets the spacing wax in a smooth joint. In order to duplicate the master model, a vinyl-polysiloxane silicone placed in a flask is used. After its setting, the duplicate model is poured (Figure 10), using class IV hard plaster.

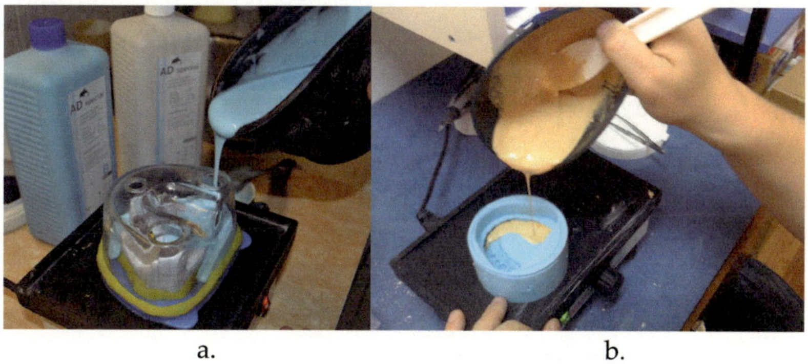

Figure 10. (a) Duplication of the model (b) Casting of the duplicate model using class IV hard plaster

The elements of removable partial denture's wax pattern are as follows (Figure 11a): the main connector, made of red wax (so that its thickness is twice as normal), the saddles and the Ackers circular clasps, made of blue wax. Injection bars are also required for those areas of the framework that are not visible in the finite piece. A large central shaft is necessary in order to connect with the main connector, through which the initial injection takes place. Unlike the pattern of a metallic framework, the patterns of the metal-free framework have to be 50% thicker (clasps, occlusal rests and main connector) [12, 13].

Spruing the framework is performed using minor sprues of 2.5 mm calibrated wax connected to one major sprue (Figure 11a).

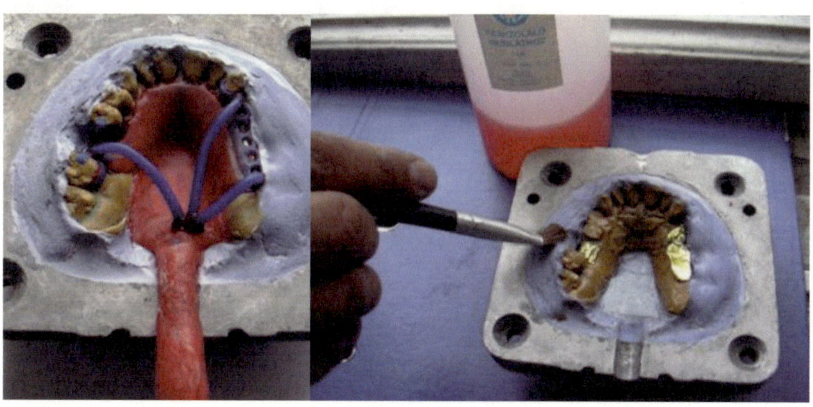

Figure 11. (a) Wrapping the wax pattern frame of the removable partial denture (b) Insulation of the investment

Surface tension reducing solution is applied and the wax pattern is then invested in a vaseline insulated aluminum flask (Figure 11b). Class III hard stone is used as an investment. The gypsum paste is poured into one of the two halves of the flask and the duplicated model containing the framework pattern with sprues attached is centrally dipped base-face down. After setting, the gypsum surface is insulated and the second half of the flask is assembled. Class III hard stone is once more prepared and the flask is submerged in warm water in a thermostatic container. The two halves of the flask are disassembled and the wax is boiled out using clean hot water. The surface of the mold is then insulated and treated with a light-curing transparent varnish in order to obtain a shining aspect [12, 13].

Injection is carried out with the R-3C (Flexite) injector (Figure 4b), which does not take up much space as it can be mounted on a wall as well. The device has the following parameters: digital control, preset programs for different types of thermoplastic resins and programs that can be individually set by the user. The pressure developed is 6-8 Bar [23].

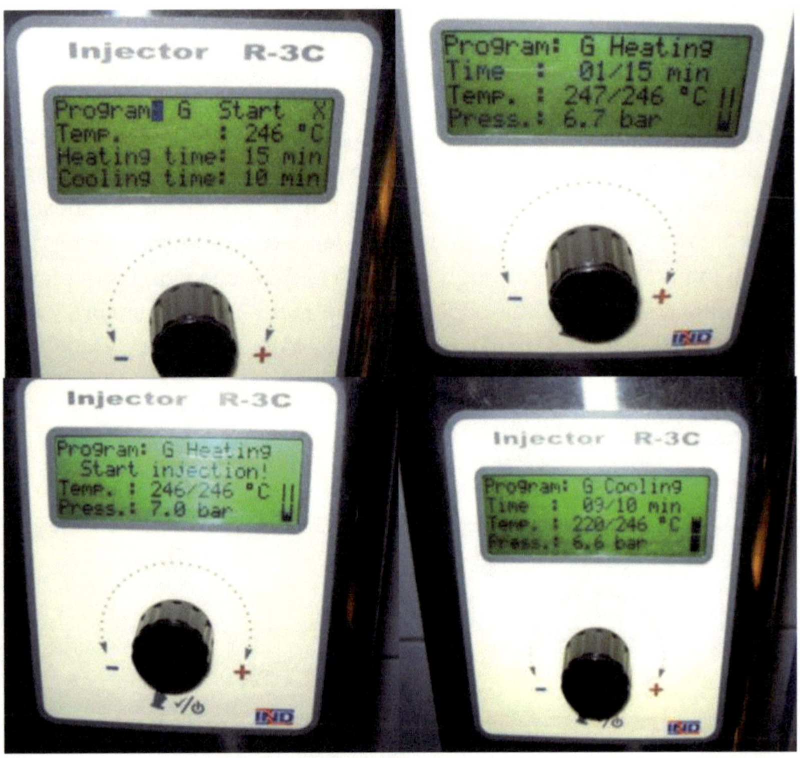

Figure 12. Schedule of "G" program of injecting the thermoplastic material (a) Start (b) Heating (c) Injecting (d) Cooling

Before starting injection, the valves of carbon dioxide tank are checked to make sure the injecting pressure is according to procedure demands (7.2-7.5 Bar). Preheating temperature and time are also checked (15 minutes at 220°C). The selected cartridge (quantity and color) is introduced into one of the two heating cylinders and the preheating process is then activated (Figure 12). After preheating ends, the two halves of the flask are assembled and fastened. Early assembling of the flask is not indicated because water vapor condensation might occur inside the mold, with negative effects on the quality of the injected material. The flask is inserted and secured in the injecting unit.

The injection process takes only 0.25 seconds and it is initiated by pressing the key on the control panel. The pressure is automatically kept constant for one minute so that setting contraction is compensated. This stage is indicated with the sign "----" on the screen. The cartridge is separated and the flask is then released and pulled out. In order to achieve optimal quality of the material, the flask is left to cool slowly for 8 hours [12, 13].

Before investment removal, screws are loosened and the flask is gently disassembled (Figure 13).

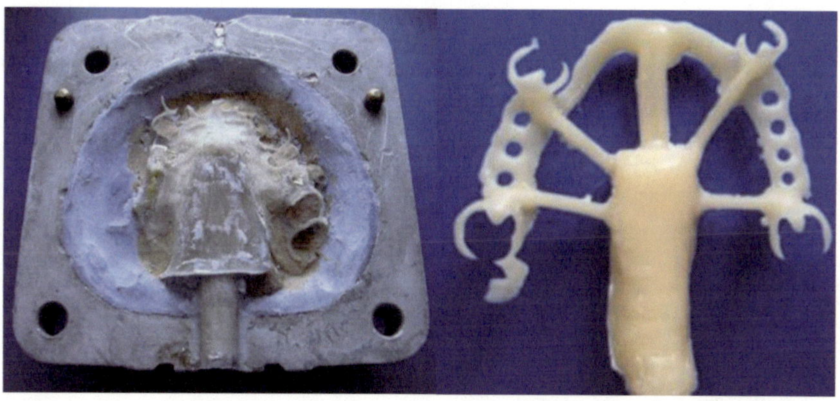

a. b.

Figure 13. Disassembling the framework of the acetal resin removable partial denture (a) The framework is still in the flask (b) Disassembling is complete

The sprues are cut off using low-pressure carbide and diamond burs to avoid overheating the material. Finishing and polishing is performed using soft brushes, ragwheel and polishing paste (Figure 14).

Disassembling the framework of the future removable partial denture is followed by matching it to the model, processing and finishing this component of the removable partial denture (Figure 15).

Once the framework is ready, the artificial teeth are set up. Wax patterns of the saddles are constructed by dropping pink wax over the framework. Teeth set up starts with the most mesial tooth, which is polished until it esthetically fits onto the arch [24].

After properly setting of all the teeth are the wax pattern is invested in order to obtain the acrylic saddles. An impression of the wax pattern placed on the master model is made by using a putty condensation silicone.

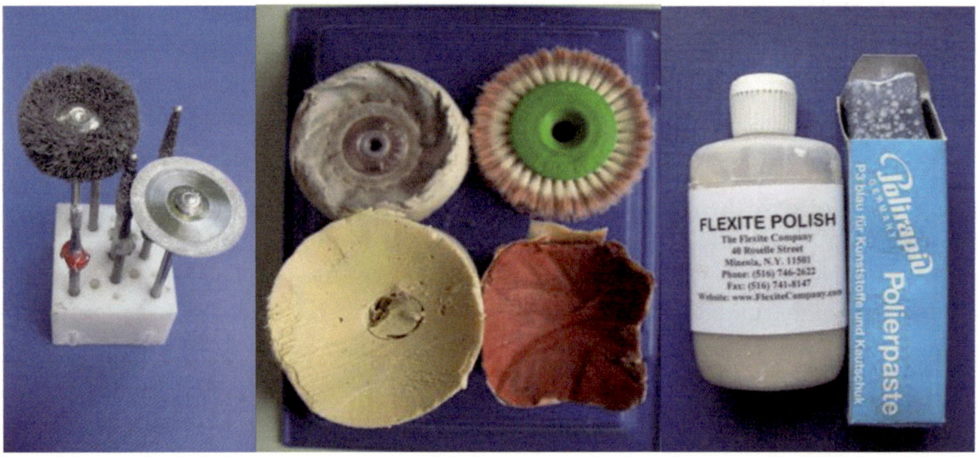

a. b. c.

Figure 14. (a) Tools used for processing the acetal framework (b) Tools used for finishing and polishing the acetal framework (c) Special polishing paste

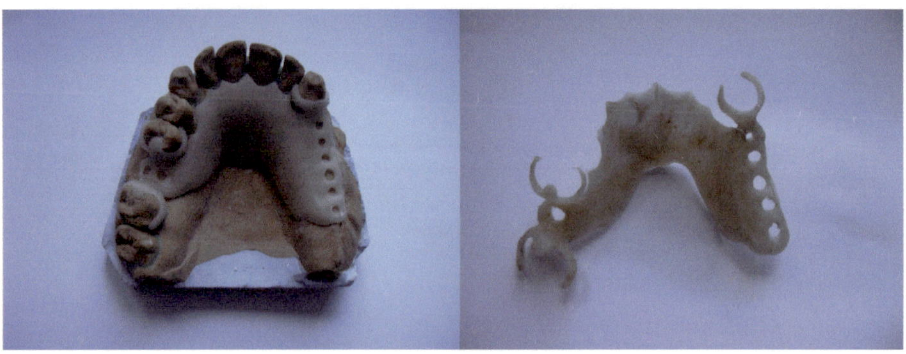

a. b.

Figure 15. (a) Matching the acetal framework to the model (b) The finished acetal framework

After setting, impression is detached, wax is removed, and the teeth, framework and the master model are thoroughly cleaned. Openings are being cut on the lateral sides of the impression and the teeth are set in the corresponding places inside the impression [13]. After insulating the master model, the framework is placed and the impression set in its original place. The acrylic component of the denture is wrapped as usual, using rectangular flasks and a class II plaster (Figure 16a).

Self-curing acrylic resin is prepared and poured inside the impression through the lateral openings. The cast is introduced into a heat-pressure-curing unit setting a temperature of 50°C and a pressure of 6 bars for 10 minutes to avoid bubble development. Once the resin is cured, the impression is removed [12, 13]. Burs, brushes, ragwheels and pumice are used to remove the excess, to polish and finish the removable partial denture (Figure 16b).

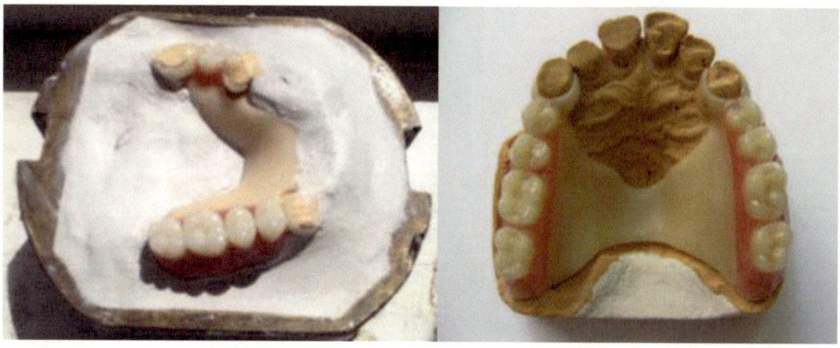

a. b.

Figure 16. (a) Wrapped wax pattern with teeth (b) Partial dentures made of acetal resin and acrylic resin

The result is a consistent removable partial denture with no macroscopic deficiency even in the thinnest 0.3-0.5 mm areas of clasps, which means the technology is effective.

3.2. Removable partial dentures made of different types of polyamides

Making polyamide resin removable partial dentures does not require so many intermediary steps as those made of acetal resins. The steps are similar to those followed for acrylic dentures, but with thermoplastic materials the injecting procedure is used. The clasps are made of the same material as the denture base, when using superflexible polyamide or ready-made clasps, in the case of using medium-low flexibility polyamide [12] (Figure 17).

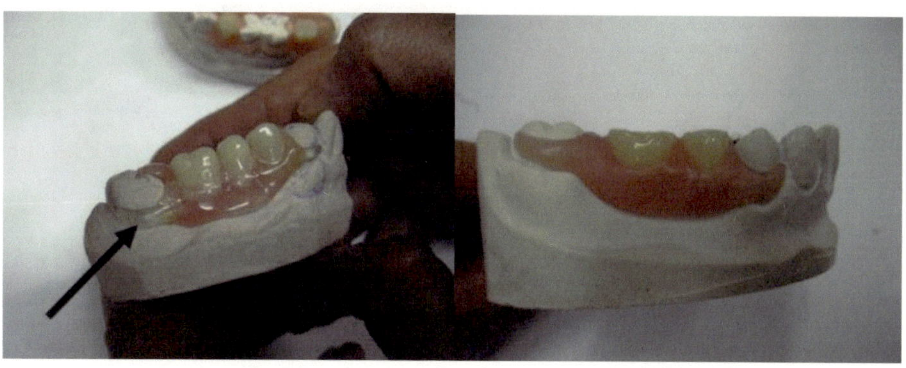

a. b.

Figure 17. Polyamide removable partial denture with (a) pre-formed clasps (b) clasps made of the same material as the denture base

Using flexible polyamide is indicated in cases of retentive dental fields (Figure 18).

When manufacturing polyamidic dentures, the support elements blend in with the rest of the denture, as they are made of the same material [25, 26].

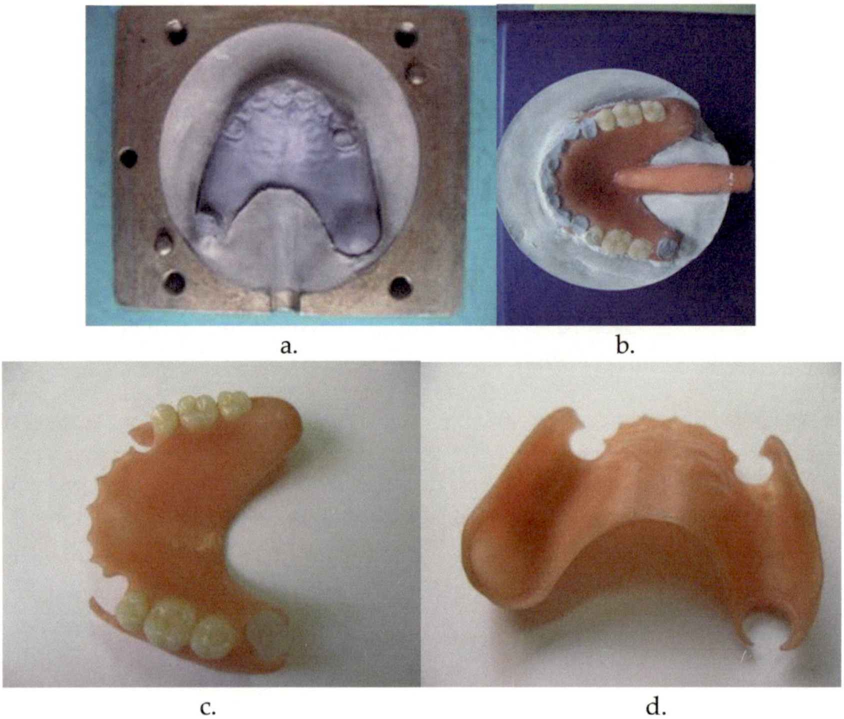

Figure 18. Removable partial dentures made of a super-flexible polyamide (a) The model with retentive tuberosity embedded in the flask (b) The denture immediately after unwrapping (c, d) The flexible removable partial denture

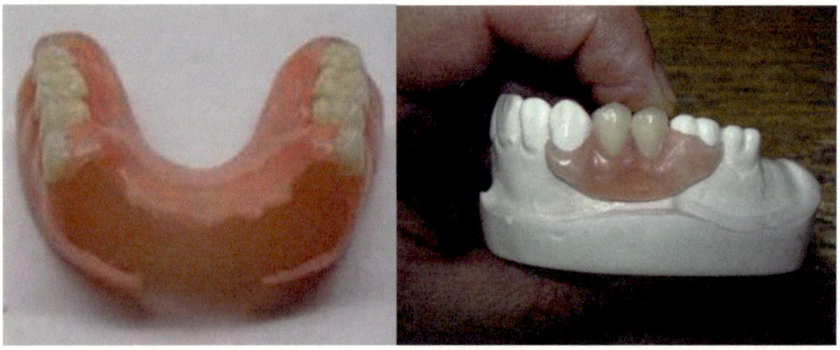

Figure 19. Medium-low flexibility thermoplastic polyamide denture

The superflexible polyamide resin is extremely elastic, virtually unbreakable, monomer-free, lightweight and impervious to oral fluids (Figures 18, 20, 21). The medium-low flexibility polyamide is a half-soft material mainly indicated for removable partial dentures. It offers superior comfort, good esthetics and no metallic taste (Figures 19 and 20). Polishing and adjusting is easy, it can be added to or relined in both dental practice and laboratory. In certain cases we used preformed clasps made of nylon. These clasps have the same composition as the polyamidic resin used for denture manufacturing, and they are heated in order to adapt (Figure 17a). This kind of clasp can be used for dentures with metal framework, or in association with injected thermoplastic resins [12, 23]. Another option we used was making the clasps of the same thermoplastic resin as the saddles or from acetal resin.

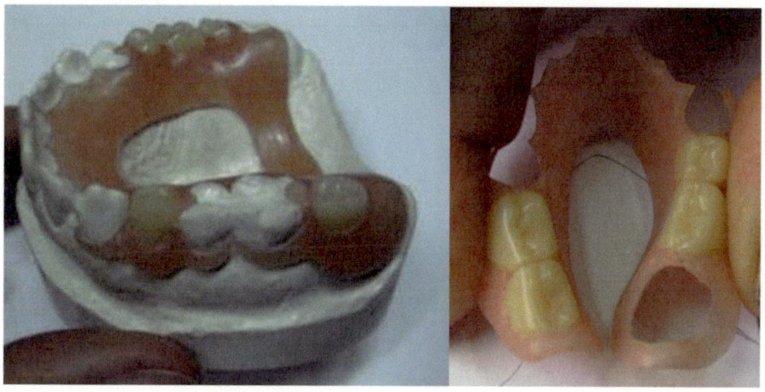

a. b.

Figure 20. (a) Medium-low flexibility polyamide partial dentures (b) Superflexible polyamide partial dentures

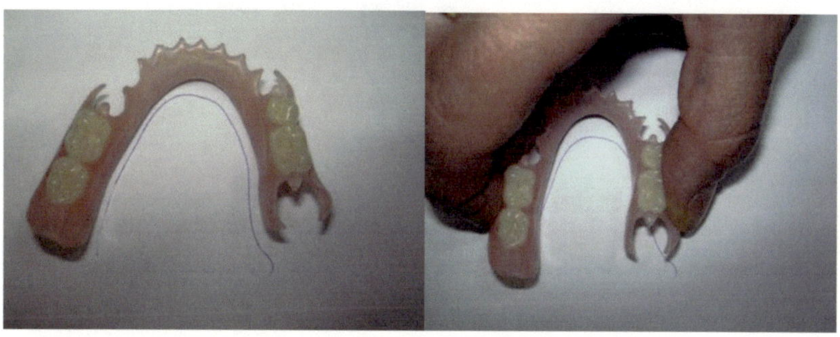

a. b.

Figure 21. (a) The superflexible polyamide denture (b) The final flexibility test

3.3. Kemeny-type removable partial dentures made of acetal

As an experiment, we managed partial reduced edentations with acetal Kemeny-type dentures (Figures 22 and 23) as an alternative to fixed partial dentures, mainly in order to test the physiognomic aspect, having the advantage of a minimal loss of hard dental substance, located only at the level of the occlusal rims, in case of posterior teeth.

Figure 22 shows wax patterning aspects and manufacturing a molar unidental Kemeny denture of acetal resin, while Figure 23 shows the way in which a frontal bidental edentation can be managed. The effectiveness of the technology is ensured by making artificial teeth of the same material.

As the material is not translucent, it is mainly suitable for dealing with lateral edentations. It can, however, be used temporarily, in the frontal area as well, in those clinical cases where short-term esthetic aspect is irrelevant [12, 23, 27].

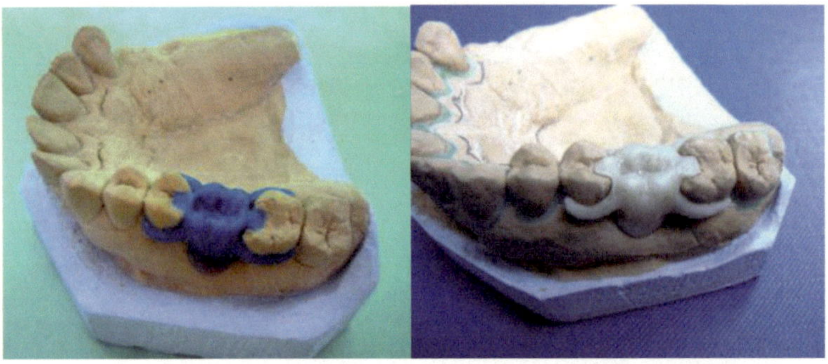

a. b.

Figure 22. Kemeny dentures: (a) Unimolar denture wax pattern (b) Denture made of acetal resin

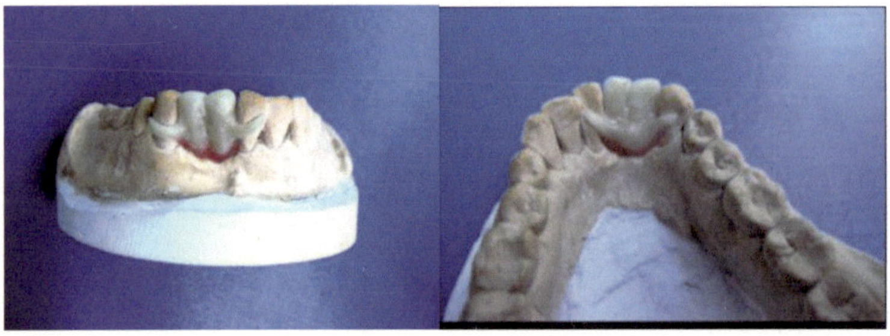

Figure 23. Kemeny-type frontal denture

3.4. Splints made of acetal resin

Thermoplastic resins are also indicated for manufacturing antisnoring devices, different types of mouthguards and splints. Parodonthotic teeth after surgery need immobilization. We experimentally manufactured acetalic resin splints (Figure 24) which turned out to be a viable solution because it matches the color of the teeth and thereby represents a temporary postoperative esthetic choice [18, 23].

3.5. Mouth guards

Mouth guards are dental appliances that can be manufactured using thermoplastic resins. The most satisfactory mouth protectors are custom-made mouth guards. This type of mouth guards is designed by the dentist. They adapt well and provide good retention and comfort. Being custom-made they interfere the least with speaking and have virtually no effect on breathing

http://dx.doi.org/10.5772/60931

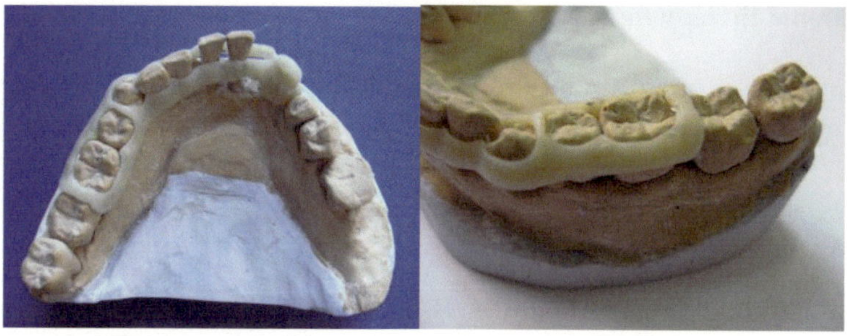

Figure 24. Thermoplastic acetal splint

[28]. Custom-made mouth guards may be classified into two types: the vacuum mouth guard and the pressure-laminated mouth guard.

The vacuum mouth guard is manufactured using a model of the upper arch. The model is casted using an impression. The thermoplastic mouth guard material, usually a polyethylene vinyl acetate (EVA) copolymer, is adapted over the model with a special vacuum machine. The vacuum mouth guard is then trimmed and polished to allow for proper tooth and gum adaptation. All posterior teeth should be covered and muscle attachments should be unimpinged. Using a vacuum machine single-layer mouth guards are manufactured. More and more, multi- ple-layer mouth guards (laboratory pressure-laminated) are preferred to the single-layer vacuum ones. The laboratory pressure-laminated mouth guard, also made from a stone cast, is a custom-made multiple-layered mouth guard that is considered the state-of-the-art mouth guard in for years. It can be made by laminating two or three layers of material to achieve the necessary thickness. Lamination is defined as the layering of mouth guard material using high heat and pressure machines. The mouth guard material should be biocompatible, have good physical properties, and last for at least 2 years [29].

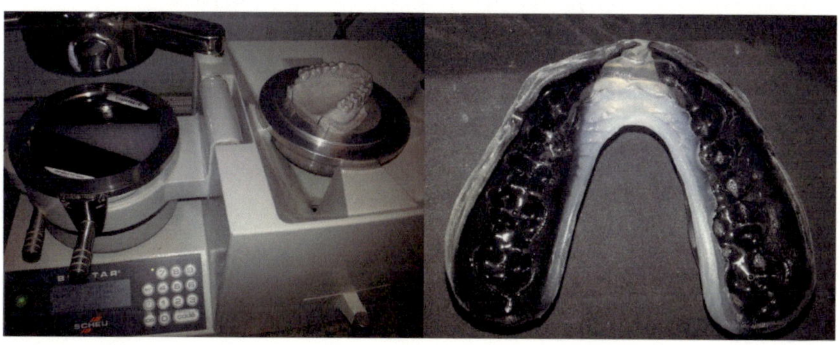

Figure 25. Laminating the custom-made mouth guard. Custom made mouth guard

We manufactured laminated custom-made mouth guards for 3 layers, with the inner layer made of a thermoplastic polyurethane which increases discoloration resistance and creates a soft inner surface feel (Figure 25).

3.6. Myofunctional therapy devices

Controlling dentofacial growth interferences is an important issue. The negative effects of mouth breathing, abnormal lip and tongue function and incorrect swallowing patterns on craniofacial development in the mixed dentition period is well known. Correcting these myofunctional habits improves craniofacial growth and decreases the severity of malocclusion [12].

Myofunctional therapy retrains the muscles of swallowing, synchronizes the swallowing movements obtaining a normal resting posture of the tongue, lips, and jaw. Myofunctional therapy may be rescheduled before, during or after orthodontic treatment [30]. The most typical age range for this type of therapy is between 8 and 16 years.

The main objective of the myofunctional appliances is to eliminate oral dysfunction and to establish muscular balance. These appliances play a certain role in orthodontics because they are simple and economical. The selection of the cases needs to be thorough and the specialist needs to be well trained in their use.

The universal size products, suitable for children between 6 and 11 years old (mixed dentition stage), allow implementing the orthodontic treatment earlier and at lower cost. These are made of a flexible thermoplastic silicone polycarbonate-urethane, a ground-breaking copolymer that combines the biocompatibility and biostability of conventional silicone elastomers with the processability and toughness of thermoplastic polycarbonate-urethanes. This type of appliances has good in vitro and in vivo stability. Its strength is comparable to traditional polycarbonate-urethanes, and the biostability is due to the silicone soft segment and end groups. Various fabrication techniques may be used in order to obtain to different. Additional surface processing after fabrication is not needed [12].

4. Errors in manufacturing thermoplastic resins dentures

Errors might occur when manufacturing thermoplastic resins dentures: the insufficient pressure at injection, which leads to lack of substance, poor polishing, or too thick saddles being some of the causes (Figure 26). These errors lead to deficiencies of the denture, which might be unusable because of esthetic deficiencies, occlusal dysmorphia, exaggerated elasticity, and decubitus areas [31].

5. Conclusions

Thermoplastics used in dentistry have known a great diversification in the last years. Processing principles are similar to the injecting technology of chemoplastics, the main difference consisting in their chemical composition, liquefying temperature of grains, injecting pressure and the fact that thermoplastic resins are monocomponent.

http://dx.doi.org/10.5772/60931

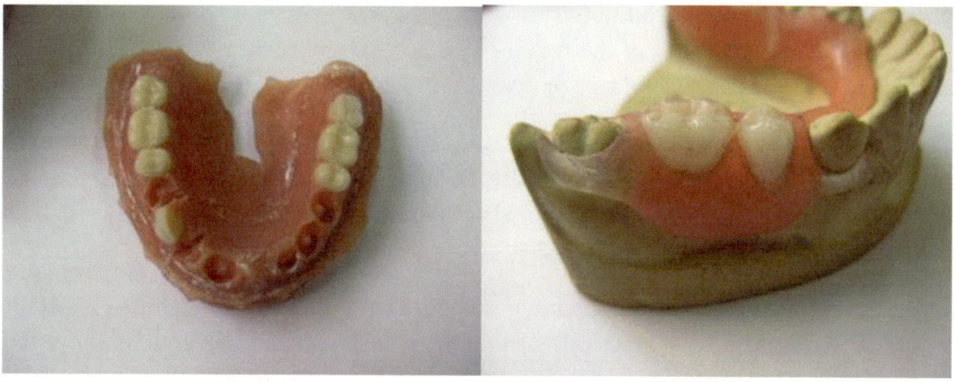

a. b.

Figure 26. Errors that might occur when manufacturing dentures from thermoplastic resins (a) Lack of substance (b) Poor polishing

Processing technology is based on the thermal plasticization of the material, in the absence of any chemical reaction. The technology of injection molding is not widely used in dental technique labs yet, as it requires special devices, but has opened new perspectives in the technology of total and partial removable dentures.

Solving partial edentations with metal-free removable partial dentures represents a modern alternative solution to classical metal framework dentures, having the advantage of being lightweight, flexible and much more comfortable for the patient. Metal-free removable partial dentures made of thermoplastic materials are biocompatible, nonirritant, sure, nontoxic, biologically inert, with superior esthetics, which make them rapidly integrate in dentomaxillary structure. They offer quality static and dynamic stability.

The clasps are made of the same material as the denture base or ready-made clasps from the same material may be used. Where the mechanical resistance of the structure comes first, the choice is an acetal resin for making the framework. Superflexible polyamide resin is especially indicated for retentive dental fields, which would normally create problems with the insertion and disinsertion of the removable partial dentures. The removable partial dentures with acetal resin frame are the most laborious to manufacture, requiring most working steps. Manufacturing the acetal framework is first, followed by the acrylic saddles and artificial teeth. A removable partial denture with an acetal resin frame is rapidly integrated into the dentomaxillary system and accepted by the patient. Such a removable denture is a comfortable solution for the partial edentulous patient, achieving the principles of static and dynamic maintenance and stability. These types of partial dentures are not bulky, the frameworks being 0.3-0.5 mm thin, and clasps are flexible and esthetic.

A particular advantage of a removable partial denture made of acetal resin applies to patients with large oral defects as a result of a maxillectomy procedure, who are due to have postoperative radiotherapy and need to have the density of the defect restored to ensure standardized radiation distribution. Different types of boluses may be used for restoration but a stent is usually needed as a support. Traditional metal-clasp retained stents are discarded in such cases as the clasps cause backscatter of the radiation beams. Acetal resin is a radiolucent material suitable for making a stent with clasps or even a removable partial denture to retain the bolus.

In the case of Kemeny-type acetalic dentures, the artificial teeth are made of the same material and in the same step as the rest of the denture. Because it is not translucent, its first indication is lateral edentations but it can be used for short periods, in the frontal area as well, if short-term esthetic aspect is not important.

Thermoplastic resins have several advantages: long-term performance, stability, resistance to deformation, resistance to wear, excellent tolerance, resistance to solvents, absence or low quantity of allergy-inducing residual monomer, and lack of porosity, thus preventing the development of microorganisms and deposits, all of which, together with maintaining size and color in time are very important characteristics, presenting a high degree of flexibility and resistance, permitting the addition of elastomers for increased elasticity or reinforcement with fiberglass, in order to increase their physical splinter quality; some of them can also be repaired or rebased.

The advantages of using the molding-injection system lay in the fact that the resin is delivered in a cartridge, thus excluding mixture errors with long-term shape stability, reduces contraction, and gives mechanical resistance to ageing.

As this class of materials, as well as the processing devices, has been continuously perfected, their future applicability in dental medicine will keep spreading.

Most probably, further chemical development of elastomeric and polymeric materials will enlarge the domain of clinical applications of thermoplastics in dentistry.

Author details

Lavinia Ardelean[1*], Cristina Maria Bortun[2], Angela Codruta Podariu[3] and Laura Cristina Rusu[1]

*Address all correspondence to: lavinia-ardelean@umft.ro

1 Department of Technology of Dental Materials and Devices in Dental Medicine, "Victor Babes" University of Medicine and Pharmacy, Timisoara, Romania

2 Department of Dentures Technology, "Victor Babes" University of Medicine and Pharmacy, Timisoara, Romania

3 Department for Preventive Dentistry, Community Dentistry and Oral Health, "Victor Babes" University of Medicine and Pharmacy, Timisoara, Romania

http://dx.doi.org/10.5772/60931

References

[1] Bortun C, Ghiban B, Sandu L, Faur N, Ghiban N, Cernescu A. Structural Investigation Concerning Mechanical Behaviour of Two Dental Acrylic Resins. Revista de Materiale Plastice 2008; 45(4) 362-366.

[2] Bortun C, Cernescu A, Ghiban N, Faur N, Ghiban B, Gombos O, Podariu AC. Durability Evaluation of Complete Dentures Realized with "Eclipse Prosthetic Resin System". Revista de Materiale Plastice 2010;47(4) 457-460.

[3] Faur N, Bortun C, Marsavina L, Cernescu A, Gombos O. Durability Studies for complete Dentures. Key Engineering Materials 2010;417-418: 725-728.

[4] Ghiban N, Bortun CM, Bordeasu I, Ghiban B, Faur N, Cernescu A, Hanganu SC. Evaluation of Mechanical Properties by Stereo-and Scanning Electron Microscopy of Some Heat Curing Dental Resins. Revista de Materiale Plastice 2010;47(2) 240-243.

[5] Phoenix RD, Mansueto MA, Ackerman NA, Jones, RE. Evaluation of Mechanical and Thermal Properties of Commonly Used Denture Base Resins. Journal of Prosthodontics 2004;13(1) 17-27.

[6] Hiromori K, Fugii K, Inoue K. Viscoelastic Properties of Denture Base Resins Obtained by Underwater Test. Journal of Oral Rehabilitation 2000;27(6) 522-31.

[7] Negrutiu M, Bratu D, Romanu M, et al. Polymers Used in Technology of Removable Dentures. Revista Nationala de Stomatologie 2001;4(1) 30-41.

[8] Reclaru L, Ardelean L, Rusu L. Toxic Materials, Allergens and Mutagens and their Impact on the Dental Field. Medicine in evolution 2008;14(3) 98-102.

[9] Rusu L, Reclaru L, Ardelean L, Podariu AC. Common Contact Allergens in Dental Materials. Medicine in evolution 2010;16(1) 15-18.

[10] Ardelean L, Rusu LC. Materiale, Instrumente si Aparate in Laboratorul de Tehnica Dentara. Timisoara: Eurostampa; 2013

[11] Bortun CM, Cernescu A, Ardelean L. Mechanical Properties of Some Dental Resins in Wet and Dry Conditions. Revista Materiale Plastice 2012;49(1) 5- 8.

[12] Ardelean L, Bortun C, Podariu A, Rusu L. Manufacture of Different Types of Thermoplastic. In: El-Sonbati AZ.(ed) Thermoplastic-Composite Materials. Rjeka: InTech; 2012. pp. 25-48.

[13] Bortun C, Lakatos S, Sandu L, Negrutiu M, Ardelean L. Metal-free Removable Partial Dentures Made of Thermoplastic Materials. Timisoara Medical Journal. 2006;56(1) 80-88.

[14] Sykes LM, Dullabh HD, Sukha AK. Use of Technopolymer Clasps in Prostheses for Patients Due to Have Radiation Therapy. South African Dental Journal 2002;57(1) 29-32.

[15] Arikan A, Ozkan YK, Arda T, Akalin B. An in vitro Investigation of Water Sorption and Solubility of Two Acetal Denture Base Materials. European Journal of Prosthodontics and Restorative Dentistry 2005;13(3) 119-122.

[16] Arda T, Arikan A. An in vitro Comparison of Retentive Force and Deformation of Acetal Resin and Cobalt-Chromium Clasps. Journal of Prosthetic Dentistry 2005;94(3) 267-274.

[17] Ozkan Y, Arikan A, Akalin B, Arda T. A Study to Assess the Colour Stability of Acetal Resins Subjected to Thermocycling. European Journal of Prosthodontics and Restorative Dentistry 2005;13(1) 10-14.

[18] Ardelean L, Bortun C, Motoc M, Rusu L. Alternative Technologies for Dentures Manufacturing Using Different Types of Resins. Revista de Materiale Plastice 2010;(47)4 433-435.

[19] Donovan T, Cho GC. Esthetic Considerations with Removable Partial Dentures. Journal of California Dental Association 2003;31(7) 551-557.

[20] Bhola R, Bhola SM, Liang H, Mishra B. Biocompatible Denture Polymers-A Review. Trends in Biomaterials and Artificial Organs 2010;23(3) 126-136.

[21] Ardelean L, Bortun C, Podariu AC, Rusu LC. Some Alternatives for Classic Thermopolymerisable Acrylic Dentures. Revista de Materiale Plastice 2012;49(1) 30- 33.

[22] Negrutiu M, Sinescu C, Romanu M, Pop D, Lakatos S. Thermoplastic Resins for Flexible Framework Removable Partial Dentures. Timisoara Medical Journal 2005;55(3) 295-299.

[23] Ardelean L, Bortun C. Metal-free Removable Partial Dentures Made of Thermoplastic Acetal and Polyamide Resins. Medicine in Evolution 2007;13(4) 18-24.

[24] Chu CH, Chow TW. Esthetic Designs of Removable Partial Dentures. General Dentistry 2003;51(4) 322-324.

[25] Szalina LA. Tehnologia executarii protezelor termoplastice Flexite. Dentis 2005;4(3-4) 36.

[26] Ardelean L, Bortun C, Podariu AC, Rusu LC. Some Alternatives for Classic Thermopolymerisable Acrylic Dentures. Revista Materiale Plastice 2012;49(1) 30- 33.

[27] Ardelean L, Bortun C, Motoc M. Metal-free removable partial dentures made of a thermoplastic acetal resin and two polyamide resins. Revista de Materiale Plastice 2007;44(4) 345-348.

[28] Ardelean L. Materiale dentare pentru tehnicienii dentari. Timisoara: Mirton; 2003.

[29] Mouthguard-infomed dental. http://www.infomed.es/seod/mouthguard.htm (accessed 22 January 2015)

[30] Quadrelli C, Gheorghiu M, Marchetti C, Ghiglione V. Early myofunctional approach to skeletal Class II. Mondo Ortodontico 2002;27(2) 109-122.

[31] Ardelean L, Bortun C, Motoc M, Rusu L, Motoc A. Errors in full denture casting using acrylic resins. Revista de Materiale Plastice 2008;45(2) 214-216.